# TIDAL ESTUARIES: MANUAL OF SAMPLING AND ANALYTICAL PROCEDURES

# Tidal Estuaries: Manual of Sampling and Analytical Procedures

**KEES J.M. KRAMER**
*Laboratory for Applied Marine Research, TNO Institute of Environmental Sciences, Den Helder, Netherlands*

**UWE H. BROCKMANN**
*Institute for Biochemistry and Food Chemistry, Hamburg, Germany*

**RICHARD M. WARWICK**
*Plymouth Marine Laboratory, Plymouth, UK*

*Published for the European Commission by*

A.A.BALKEMA/ROTTERDAM/BROOKFIELD/1994

Lay-out and realization: Electronic Publishing Centre, TNO Institute of Environmental Sciences, Delft, Netherlands

Cover photograph by Kees Kramer

LEGAL NOTICE
Neither the European Commission nor any person acting on the behalf of the Commission is responsible for the use which might be made of the following information.

Publication no. EUR 15735 EN of the European Commission, Dissemination of Scientific and Technical Knowledge Unit, Directorate-General Telecommunications, Information Market and Exploitation of Research, Luxembourg.

Published by
A.A. Balkema, P.O. Box 1675, 3000 BR Rotterdam, Netherlands (Fax: +31.10.4135947)
A.A. Balkema Publishers, Old Post Road, Brookfield, VT 05036, USA (Fax: +1.802.276.3837)

ISBN 90 5410 601 8

# CONTENTS

# CONTRIBUTORS TO THIS MANUAL

Participants in the JEEP92 programme, the JEEP92 Arcachon, Plymouth and Faro workshops, and several others, have contributed to the contents of this manual through suggestions, discussions and advice. The authors want to express their gratitude to all for their stimulating support.

Special thanks are due to those who were actively involved in the production of this manual, through discussions and/or production of texts:

M. Austen (Plymouth, UK), G. Bachelet (Arcachon, F), J. Castel (Arcachon, F), P. Caumette (Arcachon, F), J. Davey (Plymouth, UK), M. Desprez and J.P. Ducrotoy (St. Valéry s Somme, F), V. Escaravage (Arcachon), K. Essink (Haren, NL), H. Etcheber (Bordeaux, F), N. Goosen (Yerseke, NL), Th.L. Hafkenscheid and C.E.M. Heeremans (Delft, NL), C. Heip (Yerseke, NL), P. Herman (Yerseke, NL), V.N. de Jonge (Haren, NL), M. Kendal (Plymouth, UK), J. Kromkamp (Yerseke, NL), J. Mees (Gent, B), A.W. Morris (Plymouth, UK), R. Neves (Lisboa, P), M. Tackx (Brussels, B), M. Vincx (Gent, B), J.G. Wilson (Dublin, IRL) and W. de Winter (Yerseke, NL).

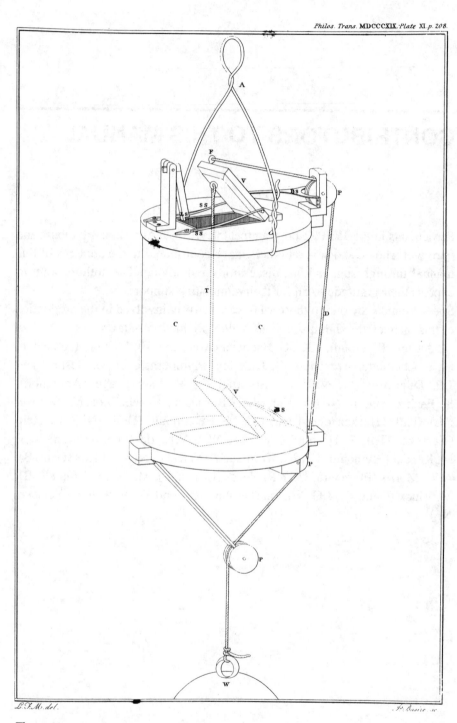

*Philos. Trans.* **MDCCCXIX.** *Plate* XI. *p. 208.*

*The past:*

*Water sampler of Marcet (From: Marcet, 1819 )*

# PROLOGUE

The EC-MAST JEEP92 Project:
Major Biological Processes in European Tidal Estuaries.

Tidal estuaries in Europe serve important economic functions including transport, industry and tourism but also drainage of wastes from domestic, industrial and agricultural activities. Estuaries are under heavy and increasing pressure from human activities in very large areas since they drain water carrying nutrients, organic matter and toxic materials from the terrestrial environment and the river system to the sea. For example, in the Dutch delta area the water of more than 300,000 $km^2$ of land surface from seven European countries is concentrated.

Whereas estuaries serve an important role for economic activities they have important natural values as well. Tidal estuaries are characterized by high secondary production rates reflected in important biomass values for benthic and zooplankton populations and the larvae and juvenile stages of fish. They serve as overwintering or passage stations for large populations of palaearctic birds. Estuarine plants and animals are adapted to high environmental variability and possess potentially valuable genetic characteristics. Salt marshes may serve as a natural defence against the rising sea level.

The major biological processes in estuaries may be linked to the production and mineralisation of organic matter. The mineralisation of the organic waste originating from human activities and the subsequent increase of nutrients and primary production is a matter of great concern. There is evidence that increased primary production is absorbed at least partially by increasing benthic metabolism and that the biomass of benthos in estuarine and coastal systems is now increasing on the long term. The eutrophication of coastal marine areas originates from land-based human activities and the effect of measures such as the reduction of phosphate and nitrate concentra-

1

tions in effluents largely depends on how these substances are transferred through the estuary, which is a non-linear system and therefore not easily predictable.

Another matter for concern is the introduction of toxic material in estuarine and coastal environments which may severely change the functioning of the estuarine filter system. Although the toxic effects of single and even mixed pollutants in laboratory conditions have been studied intensively, the fate of many pollutants in the natural estuary remains unclear. Current efforts of modelling the transfer of pollutants in the food web or measuring the accumulated toxicity in higher food web levels will be more valuable if they are based on knowledge of the important structural and functional characteristics of the estuary.

The Joint European Estuaries research Project JEEP92 is aimed at a better understanding of the effects of organic matter, which is important for the ecology of the European estuaries. By integrating results from existing European research efforts and by specific research on certain topics it is hoped to achieve a better understanding of the responses within the estuarine system. In the first programme (1991 - 1993) the major research effort was devoted to selected biological processes (primary and secondary production, bacterial production, effects or organic loading on meio- and macrofauna) as they relate to nutrients and organic matter. A second important goal was the comparison of structures and processes between different estuaries. Comparison can only be based on standardized and intercalibrated methodology and for this reason a manual representing the state of the art on chemical and biological analyses was prepared. This manual has served as the basis for the book. It is hoped that it will serve the same useful goal for the scientific community at large as it has for the scientists collaborating in JEEP92.

*Carlo Heip*
*Chairman, JEEP92*

# PREFACE

The objective of the EC sponsored JEEP92 project (Joint European Estuaries research Project) is to study the major biological processes that influence the fate of organic matter in European tidal estuaries (Elbe, Ems, Scheldt, Somme, Gironde, Shannon, and others). This involves the simultaneous determination of a selected set of state variables relevant to these major biological processes, and which involve both chemical and biological data in the water column and in the sediment. The data set will be collected by the individual participants of the project, who carry out the data collection within their respective national programmes. This means that, in principle, the participants will apply their own proven methods and methodologies within their national estuarine programmes. As the JEEP92 project combines these separate researches in an intercomparison of the different estuarine ecosystems in a mathematical model, it is essential that the data are comparable in terms of collection (sampling time and location, strategy) and analysis. It appears that data collection and interpretation are not easy tasks in the estuarine environment, mainly due to inhomogeneity and temporal variation. Under these circumstances, in general, carefully designed multi-parameter sampling programmes are necessary. Especially careful design with respect to space and time is essential.

Because of the physical and hydrodynamical differences between the various estuaries which are often highly dynamic systems, and the logistic problems that occur in estuarine research, physical, chemical and biological processes are usually only partly understood. Comparison between different estuaries may help to solve this problem.

For estuarine intercomparisons ideally only those methods of sampling and/or analysis that give the same result are to be accepted, even if the methods have a different principle of methodology. For example, even though the calculation of salinity from the analysis of chlorinity and the determination of conductivity are basically different, the final result is very

3

similar. Many methods for sampling and analysis are now standardized, sometimes for decades (*e.g.* chlorinity). Sometimes, however, several methods are in use that give non-identical results (*e.g.* DOC, POC). It seems simple to solve the problem by appointing one method as "the best" method, but it is often not clear what "the best" means. This is especially the case in sampling strategies and methods, and with new or improved analytical techniques.

Although this manual aims to establish identical methods and methodologies for application to all estuarine systems involved in the JEEP92 programme, methods and procedures presented here will be applicable to most other (tidal) estuarine systems all over the world. The authors realize that not every participant in such programmes will always have the technical means to apply each preferred method. In addition, it seems that where several methods are in use, every scientist is willing to cooperate to adopt a commonly accepted method in an international programme, but only provided that the selected method will be his own. Their is a strong need for inter-calibration of the various techniques that aim at the same result.

This manual summarizes the sampling and analytical techniques that can be applied. It proposes, in each case, one or a few methods that should preferably be applied with a view to standardization. This manual tries to make an inventory of the different options and to guide estuarine scientists towards a common use of techniques and procedures, to enhance comparability of data. It is not the intention of the authors to copy all handbooks either on physico/ chemical and biological sampling techniques in sea- or estuarine waters or on the different analyses. For detailed descriptions the reader is referred to these handbooks. Most laboratories will have at least some experience in many of the techniques described.

The use of names of manufacturers or brand names is sometimes unavoidable in a manual like this. It should be mentioned that the authors do not endorse these products.

The authors hope that, through this manual, more and more methods will become standardized throughout the estuarine project of JEEP92, and perhaps even more generally.

It was therefore decided to prepare from the original report this volume as a separate book. It is based on the JEEP92 report (Kramer *et al.*, 1992). It has been revised and enlarged. New are *e.g.* methods and procedures involving the sampling and analysis of pollutants. It was felt by the authors that these additions would improve the use of the manual.

The authors have tried to prepare this manual as clearly as possible, to ensure easy access to the information. We apologize if this has not always been successful. A manual like this will almost inevitably provoke discussions and comments. We are willing to receive any comments or corrections on the present volume, and welcome any suggestion for its further improvement.

Collection of samples is usually the basis for marine and estuarine research. The proper method of collection has always intrigued scientists, many ideas have been formulated and ingenious instruments have been designed and tested in the past. The present instrumentation in oceanography often finds its roots in these old designs, which can be over one hundred years old.
Blank pages in this volume serve as a tribute to those inventors of instrumentation and techniques. Interesting graphs have been reproduced as non numbered figures. They were copied from a collection of antiquarian books belonging to the first author, and originated from the following sources:
- Marsigli, L.-F., 1786 [1725]. Natuurkundige beschryving der zeën. 's Gravenhage
- Brunings, Chr., 1789. Verhandeling over de snelheid van stroomend water en de middelen, om dezelve op allerleïe diepten te bepaalen. In: Verhandelingen; Hollandsche Maatschappye der Weetenschappen, Part XXVI, Haarlem, pp. 1-210
- Marcet, A., 1819. On the specific gravity, and temperature of sea waters in different parts of the ocean, and in particular seas; with some account of their saline contents. Phil. Trans. Roy. Soc. (London), 109: 161-208
- Tornøe, H., 1880. On the air in sea-water. In: The Norwegian North-Atlantic Expedition 1876-1878. Chemistry part 1. Christiania, pp. 1-24
- Challenger Report, 1885. Narrative of the cruise of H.M.S. Challenger with a general account of the scientific results of the expedition, Vol. I, part 1. HMSO, London
- Thoulet, M.J., 1890. Océanographie (statique). Baudoin, Paris, pp.492
- Richard, J., 1907. L'Océanographie. Vuibert & Nony, Paris, pp. 391
- Krümmel, O., 1923. Handbuch der Ozeanographie. Teil 2. Stuttgart
- Albert 1er de Monaco, 1932. Résultats de campagnes scientifiques, Fasc. LXXXIV. Receuil des travaux publiés sur ses campagnes scientifiques, Monaco
- Rouch, J., 1943. Traité d'océanographie physique. I. Sondages. Payot, Paris, pp. 256

The JEEP92-project was sponsored by the Commission of the European Communities, Directorate General for Science, Research and Development, Marine Science and Technology Programme under contract: MAST-CT90-0024.

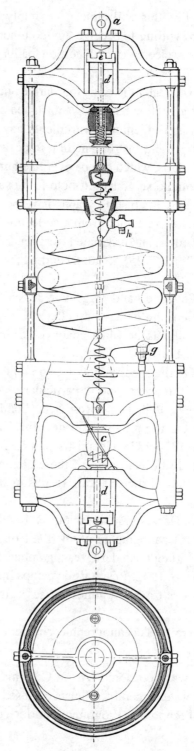

*The past:*
*Water sampler of Wille (From: Tornøe, 1880)*

# 1. SUMMARY

This "Manual" refers to a set of methodological descriptions of sampling strategies, sampling tools and methods for the study of chemical and biological variables in water and sediments of tidal estuarine environments. It is based on a report prepared for JEEP92 (Joint European Estuaries Project), of the EC-MAST programme. The objectives and approaches of this project are easily transferred to other estuarine programmes. A detailed description of both the sampling strategy and procedures, as well as the analytical methods, are presented. Comparisons between methods are made in terms of their optimal value within the estuarine research context. The sampling methods and analytical techniques aim at the study of major biological processes in estuarine systems and involve the abiotic (*e.g.* organic matter, nutrients, S, T, pH, oxygen, light, pollutants) and biotic compartments (bacteria, phyto- and zooplankton, hyperbenthos, micro- and macrophytobenthos, meio- and macrofauna). The set is based on descriptions in the literature, and involves minimum and optimum sampling schemes (frequencies and spatial distribution).

For reasons of easy comparison of results (within the JEEP92 programme), preferred sampling techniques and analytical methods are proposed.

Minimum requirements guarantee the establishment of a seasonal (pseudo)-synoptic, structural and functional comparison of tidal estuaries. Gradients and tidal variation of the ecosystems include only the main variables at some selected seasonal situations.

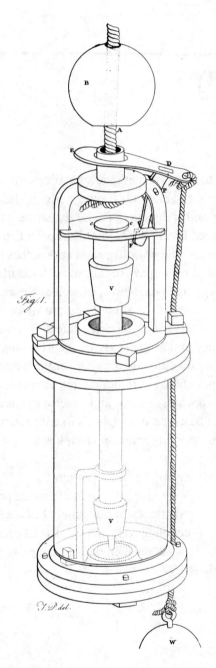

*Fig. 1.*

*The past:*
*Improved water sampler of Marcet (From: Marcet, 1819)*

# 2. INTRODUCTION

Estuaries are the areas where fresh and marine waters meet. They are consequently zones of transition, often with strong gradients and discontinuities. These are observed in the distribution of the chemical components, both major compounds like salt and suspended particulate matter (SPM or seston), and minor constituents, for example nutrients, organic matter, oxygen, etc. Estuaries are also highly dynamic zones where, for example, tidal effects exert a major influence on the influx and efflux of materials and the transport of dissolved and particulate materials within the estuarine boundaries. High turbidity zones (turbidity maxima), formed as a result of tidal circulation processes, are almost unique for estuarine environments, and are considered to be (bio-geo)chemically highly reactive.

These conditions have consequences for the estuary as a habitat for pelagic and benthic organisms. Water quality (especially temporal and lateral salinity distribution, but also nutrient regimes, oxygen concentrations, pollutant loads etc.) and sediment quality (grain size distribution, carbon content, the stability of sediment etc.) will determine which species are present and in what (relative) amounts.

Each estuary will, because of its specific geological and geographic structure, its hydrodynamics, the amount and quality of inflowing fresh water, and many other factors, be very specific: it will in fact be unique. Due to climatological changes, interferences of the diurnal light cycle, changing tides and wind forces, etc., the specific situation in an estuary at any one time will always differ from situations measured before. Much estuarine research aims to investigate the system not only in a descriptive way, but also to define and quantify the typical processes that are biological, chemical and/or physical in nature. In theory the nature of these processes should not be limited to the estuary under consideration, but have a more universal character.

Comparison between different estuaries therefore seems a useful approach to investigating general estuarine processes. In practice it is highly preferable for the same group of scientists to collect the data in the various estuaries. Comparisons are difficult to perform if one has to rely on literature data only. In a programme where data on different biological and (geo)chemical characteristics of tidal estuaries are to be collected and compared, it is essential that similar strategies and methodologies are used, and that through intercalibration the comparability of the data is tested, especially when the resulting data are to be used in mathematical modelling.

Additionally, the development of a consistent international approach and harmonisation of methodologies may provide a training element to enhance transfer of knowledge to laboratories that are less experienced in one or more fields of inter-disciplinary interest. It will certainly stimulate a more integrated approach to the study of estuarine biogeochemical processes.

Much information on the methodology of the collection of physical data and of chemical and biological samples from the marine environment is at present available (see for example the COST 647 programme for biological sampling in coastal systems). These existing accounts are usually only relevant to specific programmes, and usually cover only a limited amount of methods and techniques, due to their specialised scope.

Additionally, estuaries provide very specific biogeochemical characteristics, that make sampling operations and procedures difficult to design. For example tidal currents, salinity, pH, oxygen, nutrients and organic matter concentrations, etc., show often steep gradients, which influence the species distribution and diversity even over small spatial scales. Tidal currents will influence the suspended matter concentration and the relative contribution of resuspended sediments. It is felt that the interpretation and comparison of many historical data sets from estuarine regions is hampered by random sampling procedures, ignoring physico-chemical variability.

In recommending the following set of procedures we have not simply advocated those which are theoretically the best but often cannot be realized. Instead we have adopted a more pragmatic approach which provides an optimum between applicability, availability and quality of the methods.

Wherever possible, internationally accepted procedures and strategies are the ones which are recommended in this manual.

This manual is based on the common experience of the authors and the participants in the JEEP92 programme and reflects the current knowledge of sampling strategies and analytical methods as found in the literature.

Examples of guidelines for the design of strategies and the performance of estuarine sampling programmes can be found for example in:
- Aminot, A. & M. Chaussepied, 1983. Manuel des analyses chimiques en milieu marin. Centre National pour l'Exploitation des Océans (CNEXO), Brest, pp. 395
- Anon., 1990. Guidelines for the sampling and analysis of trace metals in seawater under the Joint Monitoring Programme (JMP). Annex to the recommendations at the fifteenth meeting of JMG, Lisbon, 1990
- Baker, J.M. & W.J. Wolff (eds), 1987. Biological surveys of estuaries and coasts. EBSA Handbook, Cambridge Univ. Press, Cambridge, pp. 449
- Downing, J.A., 1979. Aggregation, transformation, and the design of benthos sampling programs. Can. J. Fish. aquat. Sci. 36: 1454-1463
- Jeffrey, D.W., J.G. Wilson, C.R. Harris & D.L. Tomlinson, 1985. A manual for the evaluation of estuarine quality, 2nd ed. Univ. of Dublin, pp. 161
- McIntyre, A.D., J.M. Elliot & D.V. Ellis, 1984. Introduction: design of sampling programmes. Methods for the study of marine benthos. N.A. Holme & A.D. McIntyre (eds). 2nd ed. Blackwell, Oxford, pp. 1-26
- Morris, A.W., 1990. Guidelines for monitoring estuarine waters and suspended matter. UNEP Regional Seas Programme. Reference methods for Marine Pollution Studies
- Sournia, A. (ed), 1978. Phytoplankton manual. Unesco, Paris, pp. 337
- Thistle, D. & J.W. Fleeger, 1988. Sampling strategies. Introduction to the study of meiofauna. R.P. Higgins & H. Thiel (eds). Smithsonian Inst. Press, Washington DC, pp. 126-133

More general works dealing with estuarine research are, for example:
- Aston, S.R., 1981. Estuarine chemistry. In: Chemical oceanography, 2nd ed. Vol. 7, J.P. Riley & R. Chester (eds). Academic Press, London, pp. 361-440
- Burton, J.D. & P.S. Liss (eds), 1976. Estuarine chemistry. Academic Press, London
- Day, J.H. (ed), 1981. Estuarine ecology with particular reference to southern Africa. Balkema, Rotterdam, pp. 441
- Day, J.W., C.A.S. Hall, W.M. Kemp & A. Yanes-Arancibia (eds), 1989. Estuarine ecology. Wiley, New York, pp. 558
- Green, J., 1968. The biology of estuarine animals. Sidgwick & Jackson, London, pp. 401
- Head, P.C. (ed), 1985. Practical estuarine chemistry. Cambridge Univ. Press, Cambridge, pp. 337.

- Kennedy, V.S. (ed), 1984. The estuary as a filter. Academic Press, New York, pp. 510.
- Ketchum, B.H. (ed), 1983. Ecosystems of the world 26: Estuaries and enclosed seas. Elsevier, Amsterdam, pp. 500
- Lauff, G.H. (ed), 1967. Estuaries. Am. Ass. Adv. Sci., Washington, DC, pp. 757
- Morris, A.W. (ed), 1983. Practical procedures for estuarine studies. A handbook prepared by the Estuarine Ecology Group of the Institute for Marine Environmental Research. IMER, Plymouth, 262
- Nielson, B.J., A. Kuo & J. Brubaker (eds.), 1989. Estuarine circulation. Humana Press, Clifton, NJ pp. 400.
- Olausson, E. & I. Cato (eds), 1980. Chemistry and biogeochemistry of estuaries. Wiley, Chichester, pp. 452
- Perkins, E.J., 1974. The biology of estuaries and coastal waters. Academic Press, London
- Saliot, A., 1994. Cours de biogéochimie organique marine. Océans, Institut Océanographique, Paris
- Wolff, W.J., 1973. The estuary as a habitat. An analysis of data on the soft-bottom macrofauna of the estuarine area of the rivers Rhine, Meuse, and Scheldt. Zool. Verhand. no. 126, pp. 242.
- Zutic, V. (ed), 1989. Physical, chemical and biological processes in stratified estuaries. Mar. Chem. 32 (2-4), special volume, pp. 111-390.

The present manual addresses the central theme of the JEEP92 project: the study of biological processes that influence the fate of organic matter in estuaries. It will therefore be inevitable that not every possible compartment or compound will be treated.

We have set some limits to the present manual:
- the estuary is limited by its geographical borders as defined by Fairbridge (1980) (see chapter 3) which thus includes also part of the freshwater environment and part of the coastal zone. This definition is vague with respect to the seaward boundary of the system, which will usually be chosen based on the coastline morphology. It must be realized, however, that many processes that are occurring in the estuary will also occur in the adjacent coastal area, and that for a proper understanding of many geochemical estuarine processes the open-sea endmember should be known. In other words, seawater samples of a water body that never reaches the estuarine environment should preferably be included in the sampling programme.

- the description of the hydrodynamic pattern is only marginally treated; phenomena like fronts or eddies were considered beyond the scope of this manual.
- the manual will only cover the aquatic system, including the intertidal zones, but excluding the salt marshes; it should be realized, however, that salt marshes may act as an important source (or sink) of organic matter and many other compounds, and that they should be considered in this respect; inputs from the atmosphere are consequently not covered.
- groups of organisms considered important for the present volume are limited to bacteria, phytoplankton, zooplankton, micro-phytobenthos and macro-phytobenthos, meiofauna and macrofauna. For the study of hyperbenthos standard methods have been developed during the JEEP92 project. The importance of other groups, like fish, birds and mammals is recognized. However, the present programme focuses on the study of the biological processes that influence the fate of organic matter in estuaries, with special emphasis on the compartments water and sediment. It was agreed that methods of sampling and analysis for these organisms do not fall within the scope of this manual.

# 3. ESTUARIES

## 3.1 General considerations

Pritchard (1967) defined estuaries as:

*"a semi-enclosed coastal body of water which has a free connection with the open sea and within which sea water is measurably diluted with fresh water derived from land drainage".*

This definition is rather vague, and could include many coastal areas that have no direct connection with river inputs. Therefore the definition by Fairbridge (1980) seems more suitable for the present project. It also includes part of the freshwater environment:

*"an estuary is an inlet of the sea reaching into a river valley as far as the upper limit of the tidal rise, normally being divisible into three sectors: (a) a marine or lower estuary, in free connection with the open sea; (b) a middle estuary, subject to strong salt and freshwater mixing; and (c) an upper or fluvial estuary, characterized by freshwater but subject to daily tidal action."*

The geographical boundaries between these sectors are not fixed. They are subject to the (spring) tidal movements, to seasonal variability (river discharge) and may vary according to water depth (type of estuary). In figure 1 a schematic representation of the various types of estuaries is presented. The estuary types range from those with a highly stratified salt-wedge and a sharp halocline in the vertical structure to well-mixed estuaries where a vertical homogeneity is almost reached. A summary of the major types is given in table 1, together with their main features. They are described in more detail by Morris (1983). Recognition of these hydrodynamic factors

15

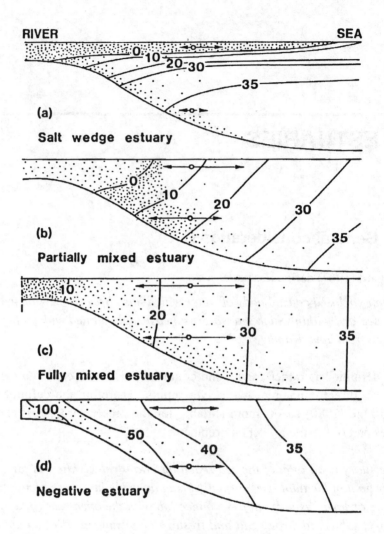

*Figure 1.*
*Schematic presentation of types of estuaries. (after Postma, 1980)*

(*i.e.* mainly river discharge, wind stress and tidal stress) will affect the design of any sampling programme.

Pelagic organisms like phyto- and zooplankton dwell to some extent with the moving waterbody during estuarine circulation. The organisms and water samples may be collected at given preset salinities. Benthic organisms, however, and the sediments, experience the sometimes large and rapid variations in salinity at given locations.

For biological purposes the *Venice classification* (Caspers, 1959) has been designed to sub-divide the estuarine system into rather complicated salinity

*Table 1.*
Estuarine types according to mixing and circulation (after Pritchard, 1955; Dyer, 1979)

| description | mixing process | vertical structure | type |
|---|---|---|---|
| highly stratified salt wedge | river flow dominates | very sharp halocline | A |
| fjord | river flow & entrainment | sharp halocline | |
| partly mixed | river flow & tidal mixing | increasing S with depth | B |
| well mixed | tidal mixing dominates | nearly vertical homogeneity | C or D (with lateral S gradient) |

*Table 2.*
Venice classification system of estuarine waters into salinity zones (Caspers, 1959)

| Venice zonation | salinity range ($10^{-3}$) | simplified zonation |
|---|---|---|
| Hyperhaline | > 40 | ) |
| Euhaline | 40 - 30 | ) Euhaline |
| (Mixo) euhaline | > 30 but < adjacent euhaline sea | ) |
| (Mixo)polyhaline | 30 - 18 | ) Polyhaline |
| (Mixo)mesohaline | 18 - 5 | ) |
| $\alpha$-Mesohaline | 18 - 10 | ) Mesohaline |
| $\beta$-Mesohaline | 10 - 5 | ) |
| (Mixo)oligohaline | 5 - 0.5 | ) |
| $\alpha$-Oligohaline | 5 - 3 | ) Oligohaline |
| $\beta$-Oligohaline | 3 - 0.5 | ) |
| Freshwater | < 0.5 | ) Freshwater |

zones. Table 2 presents the full Venice classification. For practical purposes a simplified zonation is introduced for use in this manual.

This leads to the simplified scheme (in $10^{-3}$) of freshwater (S < 0.5), oligohaline (0.5 < S < 5), mesohaline (5 < S < 18), polyhaline (18 < S < 30) and euhaline (S > 30). These simplified ranges correspond well with earlier biological findings, *e.g.* by Wolff (1973).

## 3.2 Characteristics of specific estuaries

Important features that characterize estuaries include the morphology, catchment area, river discharge and residence time (for various periods of the year), water volume, tidal amplitude, and general bathymetry. A description of the salinity distribution and the location of the turbidity maximum is required as well. In addition a characterization of the ecology of the estuary in question, subdivided into salinity related sub-regions should be given. The various groups of organisms are to be described in terms of total biomass and production, and should be completed by a short description.

Within the JEEP92 programme, an "Inventory of the physical, chemical and biological characteristics of the major European tidal estuaries" has been published as JEEP92 report (De Winter, 1992b). In Annex I a synopsis of this inventory is given which will be suitable in other areas as well. It is subdivided into several sections:

-   general description of the estuary, with its tributaries, the countries that interfere with the rivers and or estuaries, human activities, etc.;
-   economic use, especially related to industrial zones, (major) cities, harbour activities, fisheries, aqua-culture;
-   hydrology, emphasizing the catchment area, river discharge, surface area and water volume, residence time, tidal amplitude and location of turbidity maximum, if applicable at different times of the year and at various locations;
-   salinity zones, as defined by the biologically important sub-divisions (simplified Venice classification, see section 3.1): freshwater, oligohaline, mesohaline, polyhaline and euhaline. For each salinity zone the average surface area, length & width of the area, the water volume and the residual current should be given;
-   bathymetry for the same salinity based sub-areas, where surface areas are subdivided according to water depth;
-   biology, where each representative group is presented in terms of biomass

and production, again for the different salinity based zones;
-   physico-chemical parameters, indicating the most relevant variables.

Where applicable and available, additional maps (*e.g.* catchment area, estuary, sediment grain-size distribution), transects or time series (like river discharge, seston concentration, salinity, nutrient profiles) add to the value of
the synopsis. Additionally information concerning pollutant related characteristics can be added.

Clearly a complete data set will not be available for all estuaries. The
JEEP92 project may fill most important gaps for the estuaries included in the
project, however. For further details, the reader is referred to the synopsis
(De Winter, 1992b).

# 4. SAMPLING STRATEGY

The objective of sampling is to collect a portion of material from an environmental compartment (either water, sediment or biota) small enough in volume to be transported conveniently and handled in the laboratory, while still accurately representing the part of the environment sampled.

This implies that the relative proportions or concentrations of the components of interest will be the same in the samples when they are being analyzed, as they were originally in the environment.

This is a most difficult task, however. Any system, and estuaries in particular, will differ from each other and in time. It seems impossible to set guidelines that are relevant to each system. It is certainly not the intention of the authors to provide a very strict scheme of obligatory procedures. The individual scientific knowledge of the system, the specific conditions at the location and day, will largely determine the sampling programme. But, as already mentioned in the preface, some **minimum rules** have to be agreed upon if a comparison of such different systems is the objective. The minimum rules will in general not conflict with already existing programmes.

Several papers and books discuss and advise on sampling strategies in estuarine waters (Anon, 1990; Cairns & Pratt, 1986; Downing, 1989; Jeffrey *et al.*, 1985; McIntyre *et al.*, 1984; Morris, 1983, 1985, 1990; Thistle & Fleeger, 1988; Venrick, 1978). The scientific advice about the actual sampling in these references usually stops at the point where decisions have to be made for specific site selection, core diameter, etc., as different studies have different objectives and it is often considered an interference with individual scientific freedom.

Within, for example the JEEP92 programme which aims at comparison, **standardization** (to some extent) is considered essential. JEEP92 aims at a standardized minimum set of data, sampled and analysed as much as possible according to preset rules. This almost automatically implies that no complicated procedures can be proposed. Minimum sampling strategies are given here, that are taken as an example from the JEEP92 strategy, aiming at intercomparison of different estuarine systems. It should be stressed here that sampling strategies involving more than the minimum set in space and/or time, will inevitably result in a better understanding of the estuarine system and the interrelating biogeochemical processes and dynamics. If large gradients in parameters are to be expected, like during a phytoplankton bloom event, the minimum sampling frequency will be too low, and a sampling frequency at least every one or two weeks is required to cover the environmental changes adequately. Likewise, the number of locations may be increased.

In most estuaries it should be no problem to incorporate the minimum sampling programme into existing research projects. Scientists are encouraged to adapt (expand) the minimum requirements of place and time of sampling when the changes in the environmental system, or other scientific reasons, require such a step.

A discussion on some of these major constraints follows here.

## 4.1 Compartments

Each different compartment in an estuary has specific demands for sampling:
- **water**, with the dissolved compounds (by definition all material passing through a 0.45 μm filter);
- **suspended particulate matter** (SPM) or seston, which is an assemblage of living plankton, dead remains of organisms, and inorganic particles that may partly be derived from resuspension;
- **sediment**, which may vary in grain size considerably, even over short distances, and which is usually subdivided into a clay fraction (particles < 2 μm), a silt fraction (between 2 and 63 μm) and a sand fraction (> 63 μm). Sediment is frequently moved and transported by physical action: resuspension and turbation by (wind induced) waves and currents, by fluctuating tidal motion, and by organisms (bio-turbation). Due to these transport phenomena, sorting of the particles occurs. In the tidal channels we will find the coarser material, while usually at the borders of the estuary and at the sides of the tidal flats the fine material is deposited. It should be realized that as for the SPM, the sediment consists of an organic (organisms, dead organic remains, organic coatings on particles) and an inorganic fraction (sand, silt, clay);

-   **biota**, which is actually a large set of sub-compartments, each group of
    species being a separate sub-compartment.

A major distinction should be made between the organisms that are living in
the water column and those in or on the sediment (benthos). The first group
will generally float with the moving water (bacteria, phyto-plankton, zooplankton, hyperbenthos), and may therefore be subject to moderate differences in the water quality due to mixing of the salinity gradient. The benthic
organisms on the other hand, have no or only limited possibilities to migrate,
and are thus subject to the changing water quality regime under the tidal
circulation influence (micro- and macrophytobenthos, meiofauna & macro-
zoobenthos).

Free, actively moving organisms form a group in between, which may
actively search for their optimum environmental conditions (hyper-benthos,
fish). The latter group is not included in this manual, however.

---

*JEEP92  strategy (compartments):*

In order to be able to correlate the different variables that are collected, either
from the same or from different compartments, it is considered very important that
as many variables as possible are sampled together, if possible in one sample at
the same spot and time.

Only then can the results be related to each other.

---

## 4.2 Preliminary surveys

The selection of stations and the period of sampling should be based on
knowledge of the hydrographic, chemical and biological characteristics of
the estuary. In many cases this information will, at least partly, be available
from previous studies.

If this information is not available, a preliminary survey should be performed (Anon., 1990). The hydrodynamics (variation of the salinity distribution in space and time) is an especially important aspect of the design of a
sampling programme.

For the design of an effective sampling programme, tide tables and an indication of the river discharge over the year are essential.

***JEEP92 strategy (preliminary survey):***

The description of the estuary in the preliminary survey is based on a study of salinity **profiles**, both vertically and along the axis of the estuary. It is recommended as a minimum requirement that the salinity profiles along the axis are measured just below the surface (0.5 m), at mid water depth and at 1 m from the bottom. Continuous *in situ* ST or CTD measurements are much preferred, however. The stations for such a survey should be selected over the entire salinity range at approximately equal salinity intervals.

As the understanding of the general hydrodynamics and salinity distribution is the basis for estuarine ecological modelling, some synoptic measurements could be interesting, since the local stratification can be a function of the tidal situation and also because some estuaries can present important cross-gradients. Profiles along cross-sections will be interesting when the estuary is not too large (they would take too much time) or too narrow (they would not be necessary).

For the preliminary survey, at selected locations a **tidal cycle** should be recorded to estimate the biologically important minimum and maximum salinity and temperature at that location. From three or four such tidal stations the salinity ranges for the other potential sampling locations can be estimated. At 2-4 fixed stations the water velocities should be incorporated in the measurement programme.

At least two such surveys should be performed, one at maximum river discharge, and one around minimal river discharge.

## 4.3  Tidal effects

Tidal movement is one of the major driving forces in the (tidal) estuarine circulation. The most important movement in the estuary is usually along the horizontal axis. Particular attention should be paid to the difference in bottom and surface currents, particularly in salt-wedge type estuaries. Other water movements depend on the geography of the estuary, the river discharge and wind stress. The tidal movement will not only be visible in currents, but also in various major variables like salinity, turbidity, etc. (Fanger *et al.*, 1986). The currents will reach maximum speeds in either direction at mid-tide and a minimum of zero at about the change of the tides. This will affect the resuspension/deposition of the sediment and also the transport of pelagic organisms. Due to the tidal effects a more or less sinusoidal change in many constituents occurs in most estuaries. For example, a comparison between the concentration of chlorophyll *a* or of SPM will be hampered when the samples are taken at random periods during the tidal cycle, and the effects of resuspension/deposition are not accounted for.

On a daily scale, tidal effects are almost exclusively due to the M2 tidal cycle, a full tidal period covering just under 12.5 hours. This means that the tidal variation can be followed almost twice a day, which enables the additional monitoring of the diurnal cycle, the day-night rhythm, which is not a tidal but a biological cycle. In some areas deviations may occur.

The spring-neap tide cycle, active on a fortnightly scale, often has an effect upon the concentration of the constituents in the water column. Large variations in suspended matter load have been observed, for example in the Tagus estuary between spring and neap tide situations (Vale & Sundby, 1987). The fluctuations caused by these events seem important enough to take into account.

Sampling of such complicated dynamic systems requires more or less complex sampling strategies. These strategies often involve the use of a large logistic input, which is usually not realistic within limited programmes.

Synoptic sampling which means that all samples are collected at the same time involving as many sampling teams as stations, or sampling at the same (*e.g.*) mid-tide period at all locations is the best strategy but will only be possible in a limited number of occasions within the logistic scope of the programme.

A second best strategy is when all samples are collected at the same phase of the local tide, preferably around a mid-tide period (half way between the turn of the tides). This period should be calculated accurately for each location from the tide tables. Tide tables only provide information for certain defined locations and gravity influences only. However, deviations can be significant even over a small spatial scale, or under only moderately changing wind pressure. Therefore, if the actual sampling locations are not mentioned in the tide tables, which will usually be the case, it is necessary to adjust for the difference, either by models or by measurements.

As a less reliable option, one may also decide *ad-hoc* by salinity or current measurements, the tide phase related sampling time. If logistic means make it impossible to collect samples at the preferred period of the tidal cycle, the time (hours) after low water should be recorded.

There may be a difference between the mid-tides during the flood and during
the ebb (*e.g.* in current speed), but unless these differences reach unaccept-
able levels, they are, within the scope of most programmes, neglected.

An alternative but less favourable approach involves sampling at the turn of
the tides, either at high or low water, but the measured situation is certainly
not representative of the average situation in the estuary.

For routine analysis, to avoid the dominating effects of spring-neap tide
cycles, the samples should be collected in the (mid) period in between the
spring tide and the neap tide, and on a daily scale in the mid-period in between the turn of the tides. This can be considered as some kind of average
situation.

Additional sampling efforts to cover daily, diurnal and fortnightly processes
is highly recommended. Therefore, at least one double tidal period should be
covered, and preferably also the spring and the neap tidal situation. This
should be combined with the sampling for river discharge phenomena.

## 4.4 Residence time, river discharge

The concept of residence time of an element is defined as: the ratio of the
amount of a variable (dissolved or particulate) in a given environmental
compartment (water, sediment or biota) over the amount of the compound
supplied (or removed) per unit time:

$$\tau = A/(dA/dt)$$

where A is the total amount of the element in suspension or solution, and
dA/dt is the input to the system (Barth, 1952). River discharge will be the
most important driving force for the residence time.

Residence times vary between estuaries and within estuaries at different locations and during different periods of the year (seasonal variations), but
also within the seasons major variations occur (see also De Winter, 1992b).
Long term averages sometimes do not show the large variations that would
be expected and may therefore not be used for data interpretation.

In some estuaries, the residence time is in the order of days (*e.g.* Rhine), and
in others it may reach several months (Scheldt, Tagus). A water column
sampling programme that takes several days will inevitably result in a set of
samples that have no or only limited relation to each other. The same water

mass is not sampled, but a sequence of probably different water masses. A sampling strategy that is (pseudo)-synoptic is favoured here, or a scheme that collects the water samples within 1 to 2 days. This problem is ameliorated in estuaries with a longer residence time, where a collection of samples taking several subsequent days generally presents no problems of interpretation. In the latter programmes it is advisable to collect the samples in a sequence according to the longitudinal salinity profile.

In highly stratified estuaries the surface and bottom waters may behave differently, and have completely different residence times (Dyer, 1991). In these cases straightforward residence time calculation based on inputs of fresh water only is no longer valid. A method of computing the exchange time of freshwater and marine water in a stratified estuary has been described by Legovic (1991).

There are too many interfering processes that act on a seasonal scale. To characterize the changes in estuarine processes on an seasonal/annual scale, which are related to variations in river discharge the estuary should be sampled twice, during the minimum and during the maximum river discharge. These sampling occasions should conveniently be combined with the sampling for tidal effects.

## 4.5 Density & frequency

Sampling intervals are to be chosen on the basis of the expected frequency with which changes occur. This may vary from continuous recording or sampling every 5 minutes, to several hours or more. In situations where almost no major changes are expected, sampling once or twice a year is sufficient (*e.g.* benthic sampling).

The (minimum) number of independent samples that should be collected is dependent on the time-related processes one wants to investigate and the variation(s) in the relevant components that can be expected.

The statistical treatment (and requirements) of sampling in biological systems, which can also be applied to the non-biotic sampling is given by Green (1979), while a discussion on the sample size and number of replicates is presented by Bros & Cowell (1987). The latter will in part depend on the kind of statistical analysis envisaged.

When (semi)-continuous monitoring can be performed *e.g.* using sensors (salinity, temperature, dissolved oxygen, pH etc.), the sampling frequency will inevitably be high. Continuous monitoring may facilitate the selection of sampling times or intervals, *e.g.* when a rapid change is observed, the frequency of sampling can be increased. Regular sampling intervals are possible, but not a prerequisite.

To follow the tidal effects, sampling every hour over a minimum of one entire tidal cycle (13 h) is necessary. The diurnal variations should be monitored by sampling as a minimum a double set of tidal cycles (25 hours), to cover both the tidal variations and the biologically important day-night variations. The preferred procedure is, however, to sample a time series that covers 48-72 hours, at somewhat larger sampling intervals.

To investigate the tidal effects on a fortnightly scale at least one spring tide and one following neap tide should be sampled, both over one full tidal cycle, at hourly intervals.

Seasonal changes, especially related to the river discharge, require a minimal sampling programme that covers the maximum and minimum discharge periods, which should in principle be co-ordinated with the tidal stations discussed earlier.

As an example, taken from the JEEP92 project, in table 3 the minimum frequency of routine sampling and minimum number of independent replicate samples per type of sampling activity, is indicated. It will be clear that the frequency is dependent on the expected variation in the compartment: in the water phase samples are collected at least at monthly intervals, while sediment (benthos) collection is basically carried out only two times per year: just before the start of the spawning season as a measure of survival over the winter period, and in the early autumn to estimate the recruitment (northern hemisphere, temporate regions; Essink & Beukema, 1991). Table 3 tries to combine practicality (logistics) and desired sampling frequency.

## 4.6 Sampling locations (salinity, distance)

One of the major characteristics of an estuarine system is the variation in the distribution of salinity. Many processes, either physical, chemical, or biological, are related to the salinity. It is therefore an acceptable procedure to design a sampling programme that is based on salinity rather than on location, like every 5 or 10 km. The latter method may result in a large number of samples within a limited salinity range, which do not really reflect the estuarine distribution of the compounds or biological assemblages in question.

*Table 3.*
Minimum frequency and month of sampling, together with the number of replicates (example from the JEEP92 routine analysis)

| | Months: | | | | | | | | | | | | replicates |
|---|---|---|---|---|---|---|---|---|---|---|---|---|---|
| | 1 | 2 | 3 | 4 | 5 | 6 | 7 | 8 | 9 | 10 | 11 | 12 | |
| phys-chem parameters | | | | | | | | | | | | | |
| water | * | | * | * | * | * | * | | * | | * | | 1 |
| seston | * | | * | * | * | * | * | | * | | * | | 1 |
| sediments | | | | * | | | | | * | | | | 2 |
| bacteria | * | | * | * | * | * | * | | * | | * | | 1 |
| phytoplankton | * | | * | * | * | * | * | | * | | * | | 3 |
| zooplankton | | | * | * | * | | * | | * | | | | 3 |
| hyperbenthos | | | * | * | * | | * | | * | | | | 3 |
| microphytobenthos | * | | * | * | * | * | * | | * | | * | | 3 |
| macrophytobenthos | | | | * | | | | | * | | | | 4 |
| meiofauna | | | | * | | | | | * | | | | 4 |
| macrofauna | | | | * | | | | | * | | | | 4 |

*Table 4.*
Minimal sampling locations, coupled to salinity regimes (example from the JEEP-92 routine analysis)

| | salinity $(10^{-3})$: | | | | | | | | | | | | replicates |
|---|---|---|---|---|---|---|---|---|---|---|---|---|---|
| | 0 | 3 | 6 | 9 | 12 | 15 | 18 | 21 | 24 | 27 | 30 | 33 | |
| physico-chemical parameters | | | | | | | | | | | | | |
| water | * | * | * | * | * | * | * | * | * | * | * | * | 1 |
| seston | * | * | * | * | * | * | * | * | * | * | * | * | 1 |
| sediments | | * | | | * | | | * | | | | | 2 |
| bacteria | * | * | | | * | | | * | | | * | | 3 |
| phytoplankton | * | * | | | * | | | * | | | * | | 3 |
| zooplankton | * | * | | | * | | | * | | | * | | 3 |
| hyperbenthos | * | * | | | * | | | * | | | * | | 3 |
| microphytobenthos | * | * | | | * | | | * | | | * | | 3 |
| macrophytobenthos | | * | | | * | | | * | | | | | 4 |
| meiofauna | | * | | | * | | | * | | | | | 4 |
| macrofauna | | * | | | * | | | * | | | | | 4 |

classification:
I : fresh water (S < 0.5);  IV : polyhaline (18 < S < 30);
II : oligohaline, (0.5 < S < 5);  V : euhaline (S > 30)
III : mesohaline (5 < S < 18);

However, in the upper estuary, above the salt intrusion, the method of salinity based sampling is inappropriate and here a km scale might be used instead. An additional advantage of a sampling scheme based on salinity, rather that on location, is that processes can be compared in different estuaries on the same basis.

Salinity is generally accepted as the main index of mixing of seawater with water from the major river. The possible disturbing effects of tributaries, either small rivers, streams or discharge pipes and outlets, should however be recognized.

The distribution of organisms in an estuary, both in the water phase and in the sediment, is largely determined by the (extremes in) salinity (see also section 3.1; Wolff, 1973; Perkins, 1974). Biological sampling based on salinity is therefore very appropriate, and the simplified Venice classification (see table 2) can be used for guidance.

At different locations the tidal ranges of salinity will vary. From experience and/or from surveys (section 4.2) these ranges can be estimated. A very interesting situation occurs where the salinity ranges of subsequent sampling locations overlap (high water salinity station A ≈ low water salinity station B), thus building a complete coverage of the estuarine salinity regime, with a limited number of locations (but involving tidal stations with *e.g.* 13 hourly samples) (see *e.g.* Wollast & Duinker, 1982; Duinker *et al.,* 1983). An example of such an application is given in figure 2.

Instead of this laborious but rewarding method, the simpler strategy of sampling of the water compartment, including seston and organisms, at locations that are based on the salinity distribution can be used. In this way the sampling positions are not connected to fixed locations, but to the water parameter salinity. As the water movement will not be the same each time a sampling programme is carried out, this is a better approach to cover the entire salinity regime of the estuary.

For benthic sampling the situation is slightly different, as the sediment is a sedentary compartment that will be influenced by the moving water, with its variable salinities, with tidal and seasonal time scales. It is essential to sample at least the biologically important salinity zones as defined in chapter 3.1 and table 2. For sediment sampling and sampling of the benthos the locations, once selected, should be maintained over the time series to observe trends.

For both the water sampling locations and the benthic stations an indication of the variability in the local salinity regime over a tidal period (in the dry season and during high river discharge) will facilitate future interpretation of the data.

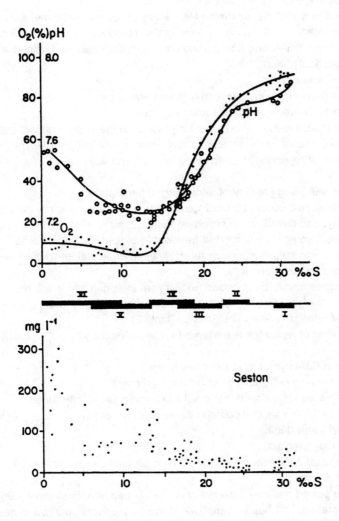

Figure 2.
Data on dissolved oxygen concentration, pH and SPM in relation to salinity,
obtained in the Scheldt estuary. The salinity limits of each of the six tidal stations
(roman numbers) are indicated (from: Duinker et al., 1982)

### JEEP92 strategy
### (tides, river discharge, locations and frequency):

The following strategy combines the *minimum* requirements for routine
sampling of the estuary, and to estimate the effects of the daily tidal processes,
diurnal variability, spring-neap tide cycles, and seasonal (river discharge)
effects.

**Routine sampling (water, including seston and organisms)**
Each sample should preferably be collected:
- at locations that are determined by salinity of the water-column, at salinities
  about every $3 \times 10^{-3}$ (table 4), thus totalling 8-10 stations; the location can
  be selected by sailing with the tide until the proper salinity is found at the
  correct tidal period,
- at mid-water depth,
- at the period half way between high and low water,
- half way between spring tide and neap tide,
- throughout the year, with higher frequencies in the spring-summer period,
  and dependent on the parameter under consideration (table 3). The
  number of replicates is given in the tables 3 and 4.

**Routine sampling (sediment, including organisms)**
Sampling should occur at 5 fixed locations that are determined by the ranges
in salinity, and based on the simplified Venice classification; in other words,
they should completely fit into the freshwater, oligohaline, mesohaline,
polyhaline and euhaline zones, preferably even under different seasonal
conditions (tables 2 and 4).
The samples should be collected twice a year in spring and in autumn.
Sometimes higher frequencies are preferred, depending on the
variable/organism under consideration (table 3).
The number of replicates is presented in the tables 3 and 4.

**Effects of tidal cycles and river discharge**
Samples should preferably be collected (minimum):
- at one sampling station that is selected within the salinity range
  $9 - 15 \times 10^{-3}$; a second option is the turbidity maximum,
- at mid water depth,
- at 1 hour intervals,
- for at least 1 full tidal cycle, 13 h (25 h is preferred to detect also the
  diurnal cycle),
- in the period half way between spring and neap tide, (additional sampling -
  tidal stations, 13 hourly samples - at two subsequent spring and neap tide
  events, is highly recommended),
- one sampling event should fall in the period of maximum river discharge, a
  second operation at the period of minimum river discharge.

## 4.7 Selection of variables

Of the long list of parameters or variables that can be studied in an estuarine system, a selection of those which are important has to be made for each programme. A priority list with a list of all variables considered in the JEEP92 project has been prepared (table 5). The priorities are an indication whether the variable is essential for the interpretation of the estuarine data set or not. Where priority 1 is indicated the variable is considered as essen-tial information that should be present in all data sets, priority 2 is a desir-able variable, while priority 3 means that the information is relevant for or can be collected by specialists only. Pollutants were not included in the JEEP92 programme; they were not considered to be essential in the func-tioning of natural ecosystems, although their effects in polluted estuaries can be substantial.

## 4.8 Sample type

Different types of samples can be collected:

a  *spot samples,* (also called catch samples), where only one sample is taken at a given location, depth and time. This will be the most applied method of sampling. Examples are a surface sample by filling a bottle, the use of a watersampler, a sediment core.

b  *composite samples.* In most cases, these samples refer to a mixture of spot samples collected at different times (or places). This method of collection reduces the number of analyses, because *e.g.* daily variations are averaged out in one analysis. If, however, the series of spot samples are not mixed but analyzed individually, information on the variability and of the analytical accuracy can be obtained, and afterwards the average might be computed. Examples are the collection of a series of sediment cores at one location, the hourly sampling over a tidal cycle, etc.

Sometimes 'time-composite' is used to distinguish from 'location-com-posite' sampling. Time-composite sampling representing a 13-hour period (one full tidal cycle) is often used, and the time interval between sampling events is 1-2 hours for many determinations.

For variables that will change after collection, and that can not be pre-served, *in situ* determinations should be applied if possible (see later). If preservatives are to be added, add them to each individual sample and not in the end to the composite sample.

*Table 5.*
Variable list, sampling and analysis-codes, symbols, units and priority status for the
JEEP92 project (pollutants were originally not involved)

**Physico-chemical variables: water (S-6.1)**

| variable | code | symbol | unit | priority 1 | 2 | 3 |
|---|---|---|---|---|---|---|
| salinity | A-7.1 | S | $10^{-3}$ | * | | |
| chlorinity | A-7.2 | Cl | $10^{-3}$ | | * | |
| temperature | A-7.3 | Temp | °C | * | | |
| light penetration (Secchi depth) | A-7.4 | - | m | * | | |
| turbidity | A-7.4 | - | % | | * | |
| dissolved oxygen | A-7.5 | $O_2$ | mg/l | * | | |
| pH | A-7.6 | - | | | * | |
| total alkalinity | A-7.7 | Alk | mmol/l | | * | |
| nutrient concentrations | | | | | | |
| nitrate | A-7.8 | $NO_3^-$ | µM | * | | |
| nitrite | A-7.9 | $NO_2^-$ | µM | * | | |
| ammonia | A-7.10 | $NH_4$ | µM | * | | |
| phosphate | A-7.11 | $PO_4^{3-}$ | µM | * | | |
| silicate | A-7.12 | $H_4SiO_4$ | µM | * | | |
| sulphide | A-7.13 | $S^{2-}$ | µM | | | * |
| sulphate | A-7.14 | $SO_4^{2-}$ | µM | | | * |
| dissolved organic carbon | A-7.15 | DOC | mg/l | * | | |
| dissolved organic nitrogen | A-7.16 | DON | mg/l | | | * |
| dissolved humic compounds | A-7.17 | - | mFl | | | * |
| dissolved total carbohydrates | A-7.18 | - | µgC/l | | | * |
| dissolved individual carbohydrates | A-7.19 | - | µgC/l | | | * |
| dissolved total amino acids | A-7.20 | - | µgN/l | | | * |
| dissolved individual amino acids | A-7.21 | - | µg/l | | | * |
| dissolved proteins | A-7.22 | - | µgN/l | | | * |
| dissolved lipids | A-7.23 | - | µgC/l | | | * |
| trace metals | A-7.24 | - | nmol/l | | | |
| PAHs | A-7.25 | - | µg/l | | | |
| PCBs | A-7.26 | - | ng/l | | | |

*Table 5 (cont).*

**Physico-chemical variables: seston (S-6.2)**

| variable | code | symbol | unit | priority 1 | 2 | 3 |
|---|---|---|---|---|---|---|
| suspended particulate matter | A-7.27 | SPM | mg/l | * | | |
| particle size (per size class) | A-7.28 | - | n*10$^3$/ml | * | | |
| pigments | | | | | | |
|    chlorophyll *a* | A-7.29 | Chl.a | mg/l | * | | |
|    chlorophyll *b* | A-7.29 | Chl.b | mg/l | * | | |
|    chlorophyll *c* | A-7.29 | Chl.c | mg/l | * | | |
|    other pigments | A-7.29 | | mg/l | | * | |
|    % degradation | A-7.29 | | % | | | * |
| particulate organic carbon | A-7.30 | POC | mg/kg | * | | |
| particulate organic nitrogen | A-7.31 | PON | mg/kg | | | * |
| particulate organic phosphorus | A-7.32 | POP | mg/kg | | | * |
| particulate total carbohydrates | A-7.18 | - | µgC/g | | | * |
| particulate individual | | | | | | |
|    carbohydrates | A-7.19 | - | µgC/g | | | * |
| particulate total amino acids | A-7.20 | - | µgN/g | | | * |
| particulate ind. amino acids | A-7.21 | - | µg/g | | | * |
| particulate proteins | A-7.22 | - | µgN/g | | | * |
| particulate lipids | A-7.23 | - | µgC/g | | | * |
| particulate trace metals | A-7.36 | - | mg/kg | | | |
| particulate PAHs | A-7.37 | - | µg/kg | | | |
| particulate PCBs | A-7.38 | - | ng/kg | | | |

*Table 5 (cont).*

**Physico-chemical variables: sediment (S-6.3)**

| variable | code | symbol | unit | priority | | |
|---|---|---|---|:-:|:-:|:-:|
| | | | | 1 | 2 | 3 |
| grain size distribution | | | | | | |
| < 2 µm, clay | A-7.33 | - | % | | | * |
| < 63 µm, silt | A-7.34 | - | % | | * | |
| 63-125 µm | A-7.35 | - | % | | | * |
| 125-250 µm | A-7.35 | - | % | | | * |
| 250-500 µm | A-7.35 | - | % | | | * |
| 500-1000 µm | A-7.35 | - | % | | | * |
| > 1000 µm | A-7.35 | - | % | | | * |
| particulate organic carbon | A-7.30 | POC | mg/kg | * | | |
| particulate organic nitrogen | A-7.31 | PON | mg/kg | | | * |
| particulate organic phosphorus | A-7.32 | POP | mg/kg | | | * |
| particulate trace metals | A-7.36 | - | mg/kg | | | |
| particulate PAHs | A-7.37 | - | µg/kg | | | |
| particulate PCBs | A-7.38 | - | ng/kg | | | |

*Table 5 (cont).*

| Biological variables: water (S-6.4 to S-6.7) | | | priority | | |
|---|---|---|---|---|---|
| variable | code | unit | 1 | 2 | 3 |
| bacteria (S-6.4) | | | | | |
| numbers | A-7.39 | $n*10^6$/ml | * | | |
| production | A-7.40 | $g/m^3$ | | | * |
| phytoplankton (S-6.5) | | | | | |
| species abundance | A-7.41 | n/ml | | * | |
| production | A-7.42 | $gC/m^3$.d | | * | |
| biomass | A-7.43 | $g/m^3$ | | * | |
| zooplankton (S-6.6) | | | | | |
| species abundance | A-7.44 | $n/m^3$ | * | | |
| stage distribution | | | | | |
| (key species only) | A-7.45 | $n/m^3$ | | * | |
| indiv. stage weights | | | | | |
| (copepods only) | A-7.46 | g | | | * |
| biomass | A-7.47 | $g/m^3$ | | | * |
| hyperbenthos (S-6.7) | | | | | |
| species abundance | A-7.48 | $n/m^3$ | * | | |
| stage distribution | A-7.49 | $n/m^3$ | | * | |
| biomass | A-7.50 | $g/m^3$ | | * | |

*Table 5 (cont).*

**Biological variables: sediment  (S-6.8 to S-6.11)**

| variable | code | unit | priority | | |
|---|---|---|---|---|---|
| | | | 1 | 2 | 3 |
| micro-phytobenthos (S-6.8) | | | | | |
| species abundance | A-7.51 | $n/m^2$ | | * | |
| production | A-7.52 | $gC/m^2.d$ | | | * |
| biomass | A-7.53 | $g/m^2$ | | * | |
| macro-phytobenthos (S-6.9) | | | | | |
| species abundance | A-7.54 | $n/m^2$ | * | | |
| biomass | A-7.55 | $g/m^2$ | * | | |
| trace metals | A-7.56 | mg/kg | | | |
| meiobenthos (S-6.10) | | | | | |
| species abundance | A-7.57 | $n/m^2$ | * | | |
| biomass | A-7.58 | $g/m^2$ | * | | |
| macro-zoobenthos (S-6.11) | | | | | |
| species abundance | A-7.59 | $n/m^2$ | * | | |
| age distribution | | | | | |
| (selected species) | A-7.60 | n/class | * | | |
| biomass | A-7.61 | $g/m^2$ | * | | |
| trace metals | A-7.62 | mg/kg | | | |
| PAHs | A-7.63 | mg/kg | | | |
| PCBs | A-7.64 | $\mu g/kg$ | | | |

c *Integrated samples*. Sometimes samples are collected at the same location but, due to horizontal or vertical variation in the composition of the estuary, they are continuously collected over (part of) a vertical or a horizontal section. Examples are sampling using a pumping system, lowering an open sample bottle, a haul with a plankton net.

To evaluate the average composition in terms of total load or mass balance, integrated samples may be collected in relation to the current velocity at that location. This method is often applied in sampling for effluent control purposes.

d *In-situ measurements*. A number of variables can be measured *in situ*, with the advantage that a direct reading may give information on the structure of the water column, or the variation with time. *In situ* measurements therefore present a valuable tool in the selection of sampling locations and times. The most important variables that can be analyzed *in situ* are salinity, pH, dissolved oxygen concentration, temperature, conductivity, turbidity and fluorescence.

For the proper interpretation of the data, and for comparison with data sets from other estuaries, a record should be made of whether the data are derived from either spot, composite, integrated or *in situ* measurements. This is therefore indicated by a special sub-code in the sampling identification code (see Annex II).

In the chapter on sampling methods (chapter 6) a distinction has been made between the various types of samples; the JEEP92 preference is thus indicated. In-situ analyses are covered in chapter 7, and Annex II.

## 4.9 Sampling depth & vertical profiles

Sampling depth, both in water and in sediment, can have a large influence on the analytical results in estuaries. The physico-chemical characteristics may change completely over 10-20 cm in the water column, and centimetres or even millimetres in the sediment. To what accuracy the sampling depth has to be known will depend on the objective of the sampling programme. In some stratified estuaries, the vertical gradient in salinity is so large that in a standard watersampler, with an internal height of ca. 0.8 m, a salinity range from $10-37 \times 10^{-3}$ can be collected (see *e.g.* in the Krka estuary, Legovic *et al.*, 1991). Such a sharp interface will not only be visible in the salinity profile, but also affect the distribution of other variables, like organic matter (Zutic & Legovic, 1987). Especially in these situations, where large S gradients exist, the salinity has to be measured in the sample that is collected for other variables, rather than derived from a combination of (CTD) salinity profile and (approximate) sampling depth (Dyer, 1991).

Each estuary, and each location within the estuary may require a different approach in sampling. For comparative purposes (for the JEEP92 programme), some minimum requirements have to be specified; more samples are, of course, never a problem.

In estuaries it is very important to have an idea of the vertical structure of the watercolumn, before any sampling starts. The use of a CTD or ST meter is almost obligatory. Vertical profiles of other variables, such as dissolved oxygen or turbidity, are helpful for the interpretation of biological, hydrodynamical and (geo)chemical processes.

From the brief preliminary survey the sampling depths can be determined.

---

***JEEP92 strategy (sampling depth, profiles):***

In principle, in case of a non-stratified estuary, a minimum of one sample should be collected at mid-water depth.

When stratification occurs, the minimum requirement is that samples are collected at the sub-surface (0.5 m, avoiding the surface micro-layer) and at 1 m above the sediment. If the water depth is more than 5 m, an additional sample has to be collected at mid water depth; above a water depth of 12 m more samples may be collected. If the vertical salinity profile indicates a strong gradient at or around one of the intended sampling depths, it is recommended that samples are collected at either side of the interface.

For physico-chemical characterization of (surface) sediments the top 2 cm should be considered as it will contain the most recent and thus relevant information. Micro-phytobenthos is sampled in the top 0.5 or 0.5 and 0.5-2 cm, meiobenthos is collected in the upper 5 cm and for macro-zoobenthos the top 25 cm should be sampled.

It does not seem possible to store all the *in situ* measured vertical profiles in a data-base where all different data are collected, and a selection has to be made. Every 1 m a characterisation should be given; where a strong gradient is observed a 0.5 m interval can be applied. For more detailed information one should always be able to check with the original data files.

For sediments vertical profiles are possible, *e.g.* using micro-electrodes, but this seems beyond the scope of the present programme.

# 5. FIELD DATA INVENTORY

During each sampling event a minimum standard set of sampling charac ter-istics should be recorded to facilitate completion of a database (see chapter 8), and to keep a record of the sampling activities. For this purpose an ex-ample Field Data Inventory is introduced here, that may be expanded if de-sired.

*Table 6.*
Estuarine sampling field data inventory ( JEEP92 )

| Institute: | Location (N): |
| Estuary: | (+E/-W): |
| Vessel: | Water depth: |
| Responsible scientist: | Sampling cruise code: |
| Year/Month/Day: | Sampling station code: |
| Time (24 h): | |

**Sampling activities (tick: )**

*\* physico-chemical parameters:*
| ☑ salinity | ☑ temperature | ☑ pH | ☐ DOC | ☐ DON |
| ☑ oxygen | ☐ nutrients | ☐ sulphide | ☐ sulphate | ☐ fluoresc. |
| ☐ Secchi | ☑ turbidity | ☐ current speed | ☐ current direction | |
| ☐ trace metals | ☐ PAHs | ☐ PCBs | ☐ other | |

*\* seston :*
| ☐ SPM | ☐ part. size | ☐ pigments | ☐ POC | ☐ PON |
| ☐ POP | ☐ trace metals | ☐ PAHs | ☐ PCBs | ☐ other |

*\* bacteria*
| ☐ numbers | ☐ production | ☐ other |

*\* phytoplankton*
| ☐ abundance | ☐ biomass | ☐ production | ☐ other |

*\* zooplankton*
| ☐ abundance | ☐ biomass | ☐ stage distr. | ☐ stage weight | ☐ other |

*\* hyperbenthos*
| ☐ abundance | ☐ stage distr. | ☐ biomass | ☐ other |

*\* sediment :*
| ☐ grain size | ☐ POC | ☐ PON | ☐ POP |
| ☐ trace metals | ☐ PAHs | ☐ PCBs | ☐ other |

*\* micro-phytobenthos*
| ☐ abundance | ☐ biomass | ☐ production | ☐ other |

*\* macro-phytobenthos*
| ☐ abundance | ☐ biomass | ☐ trace metals | ☐ other |

*\* meiobenthos*
| ☐ abundance | ☐ biomass | ☐ other |

*\* macro-zoobenthos*
| ☐ abundance | ☐ biomass | ☐ age distr. |
| ☐ trace metals | ☐ PAHs | ☐ PCBs | ☐ other |

# 6. SAMPLING METHODS

In this chapter the sampling procedures are documented to emphasize the differences that may exist between the various methods, and also the possibility or impossibility of using a given method for a specific variable. No attempt has been made to include all existing sampling methods. Methods have been selected in view of optimum applicability for the estuarine environment, keeping in mind that not all participants have access to specially designed or the newest available sampling techniques or sophisticated (automated) instruments.

All sampling procedures in this manual have been given a **S-number code** for easy reference. These numbers are used throughout the volume.

In each sampling procedure the compartment is given, and the **variables** that fall under the presented sampling method(s).

The list of sampling methods is sub-divided into different compartments and sub-compartments:

    sampling water (S-6.1)
    sampling seston (S-6.2)
    sampling sediments (S-6.3)
    sampling biota  (S-6.4 – S-6.11):
    bacteria, phytoplankton, zooplankton, hyperbenthos, micro-phyto-
    benthos, macro-phytobenthos, meiobenthos and macro-zoobenthos.

A short introduction presents the different possible sampling methods for the given compartment and variables for estuarine sampling .

Often the final results of different sampling techniques do not compare well. This can be a problem for intercomparison. In these cases the methods will have different numbers: Method #1, Method #2, etc. Unless stated otherwise, Method #1 is preferred over Method #2, etc. Where more techniques or methods are possible that give (nearly) identical results, this is indicated using suffixes: Method #1a, Method #1b.

Participants (within the JEEP92 programme) are strongly encouraged to apply the first sampling method given (thus Method #1, #1a or #1b, ...) as much as possible. The only reason why other sampling methods are treated here is to provide the possibility of including old data sets in the data base, and of separating data sets that are based on different sampling methods in the data treatment.

The number of individual replicate samples that are to be collected has been given in table 3.

*In situ* measurements, *e.g.* of salinity, temperature, turbidity, dissolved oxygen content, pH and others, are found in the chapter on analytical procedures (chapter 7).

An overview of the various sampling methods, together with their sampling codes and sub-codes for additional information in the JEEP92 data base is given in Annex II. This list can be used as a quick reference guide.

A short description of each procedure follows. References will be provided where applicable. For further details on sampling techniques, refer to the following references, handbooks and standard works, which contain valuable chapters on sampling methodology:

- Aminot, A. & M. Chaussepied, 1983. Manuel des analyses chimiques en milieu marin. Centre nationale pour l'Exploitation des Océans (CNEXO). Brest, pp. 395
- Grasshoff, K., M. Ehrhardt & K. Kremling (eds), 1983. Methods of seawater analysis. 2nd ed. Verlag Chemie, Weinheim
- Head, P.C. (ed), 1985. Practical estuarine chemistry. Cambridge Univ. Press, Cambridge , pp. 337
- Higgins, R.P. & H. Thiel, 1988. Introduction to the study of Meiofauna. Smithsonian Institution Press, Washington DC
- Holme, N.A. & A.D. McIntyre (eds), 1984. Methods for the study of marine benthos. Blackwell Sci. Publ., Oxford, pp. 387
- Morris, A.W. (ed), 1983. Practical procedures for estuarine studies. A handbook prepared by the Estuarine Ecology Group of the Institute for Marine Environmental Research. IMER, Plymouth, pp. 262
- Sournia, A. (ed), 1978. Phytoplankton manual. Unesco, Paris, pp. 337
- Tranter, D.J. & J.H. Fraser (eds), 1968. Zooplankton sampling. Unesco Monographs on oceanographic methodology, 2. Unesco, Paris, pp. 174

**S-6.1**

---

**Compartment:   water**

**Variable(s):**       **salinity, chlorinity, pH, nutrients, alkalinity, DOC,
                    trace metals, PAHs, PCBs, seston**

**Introduction:**
Sampling of water for physico-chemical analyses is one of the most common
sampling events. For estuarine work several special considerations are to be
considered (inhomogeneity, sampling depth, etc.), as discussed in chapter 4.
Possible techniques involve: sampling directly in a sample bottle, or using a
bucket, water samplers, or a pumping system. Both spot samples and integrated
samples can be obtained. When sample bottles are filled, take care to clean them,
including the caps, with water from the sample location at least twice.
The sea-surface microlayer contains elevated or even high concentrations of
many compounds. Unless the sampling is specially focused on this layer (in
which case special sampling techniques should be applied), one should avoid
collecting this layer. The sampler should pass the sea surface microlayer
preferably in closed condition (as is possible when using sample bottles as
samplers and with Go Flo samplers), or should pass the surface as quickly as
possible.
Sampling of the bottom waters may also lead to problems. A weight is often
connected below in order to be able to lower the sampling gear. When this
weight touches the bottom, resuspension of sediment will occur which could be
collected in the sampler. This will cause a misinterpretation of suspended
matter contents and of the compounds attached to it.
Most samplers are suitable for taking standard water samples intended for the
analysis of nutrients, salinity etc. (Grasshoff *et al.*, 1983). For other analyses
special samplers have been designed.
*Trace metals.* For trace metal analysis contamination control has been considered
of the utmost importance for some considerable time now and trace metals need
to be sampled using special sampling equipment. The inclusion of metal parts in
the sampler itself or the hydrowire, and rubber seals is not allowed due to the risk
of contamination. Typical materials for trace element collection and storage
involve polythene, polypropylene or the more expensive (FEP) Teflon (*e.g.*
Durst, 1979). Special, trace metal free samplers have been developed, for which
(inter)comparison of samplers has been reported (*e.g.* Spencer *et al.*, 1982;
Bewers & Windom, 1982; Wong *et al.* 1983), and optimal sampling procedures
described (*e.g.* Berman & Yeats, 1982; Yeats, 1987). Information is also
available on proven cleaning procedures (Laxen & Harrison, 1981; Berman
*et al.*, 1983). Most companies supply special samplers for trace metal work. The
collection of the water from the sampler and the following treatment is preferred
in a contamination free atmosphere (clean room or at least clean bench). In some
cases, such as the sampling of the hydrophobic organo-metal compounds, like

organo-tins (TBT, DBT, etc.) a sampler cannot be used as a result of the sorbing character of these compounds to the sampler wall (which will result in an underestimation of the concentration when collected from the sampler). In this case the collection directly in cleaned sample bottles (polycarbonate, glass or Teflon) is the only option.

Typical sampling volumes for trace metal analyses are in the order of one litre.

*Trace organics: PAHs, PCBs, etc.* In contrast to trace metals, the samplers for trace organic pollutants should be of glass, stainless steel or Teflon, for similar reasons of minimizing contamination. Due to the high risk of adsorption to sampler surfaces, most specialists prefer to collect the sample directly in glass bottles that serve as sample bottle (Duinker & Hillebrand, 1983; Hillebrand & Nolting, 1987). Due to the low concentration levels of PAHs and PCBs in seawater, relatively large volumes (up to hundreds of litres) should be collected, imposing logistic and handling problems. Contamination control should be in force whenever possible. In (polluted) estuaries one to several litres will usually suffice, depending on the compounds of interest.

**Sampling methods:**

*Method #1a. Water-sampler.* The water-sampler is much preferred as it can be operated much more reproducibly and at any depth desired. Before actually using the sampler it should be checked for proper functioning, and cleaned from shipborn dust by moving it several times up and down in the water. Mounting and operation of water-samplers is model specific, and one is referred to the individual manual.

Most samplers operate in a vertical position, and will thus collect a vertical segment of water, depending on the size of the sampler up to 1 m, that may cross boundaries between different water types. Sampling of the surface microlayer should be avoided.

The sampler is mounted to a weighted hydrowire, and closed at the desired depth by a dropping weight. In case of strong currents when the ship is at anchor, the sampler may be used sliding along a separate hydrowire that is kept almost vertical using a heavy weight or better, a type of 'depressor' as used in zooplankton catches in the open sea. In all cases the actual depth should be estimated as accurately as possible using a graduated line. Assume the middle of the sampler as the actual sampling depth and correct for deviation from the vertical.

*Method #1b. Sample bottle.* Collection of water samples direct into a sample bottle, or the use of such bottles as samplers has been used for many years. This method prevents contamination of the sample by using a sampler. In addition, samples can be collected directly in the preferred type of container (material). It is most simple to lower the bottle by hand below the surface of the water and let it fill. Sampling of the surface microlayer should be avoided. For non-surface, or better sub-surface samples, many constructions have been designed which either lower the bottle in the open position, or open it at the desired depth. The latter may be of the displacement type, that allows a steady flow into the bottle. In the open sea the method is limited because the pressure difference does not allow sampling at a depth of more than 20-30 m, depend- ing on the type of bottle (see figure 3). In estuarine waters this is usually not a limitation. Lowering in the open position collects samples not at discrete depths but some sort of integrated sample. However, collection is not even over the sampling path, because of pressure differences. An active opening procedure is therefore preferred. The bottle(s) are lowered attached to a graduated polythene line in a suitable weighted enclosure and opened at the desired depth.

*Method #2. Pump.* Pumping of water for collection has become popular for use when a ship is under way, or for large volumes of water. Several types of pumping system are used, both *in situ*, peristaltic, bellow, etc. pumps or using a vacuum behind the collection bottle. Most pumps will provide sufficient pressure, provided they do not have to suck the water to a considerable height first. Mounting of the pump close to the surface (or under water) is essential. The practical sampling depth is up to about 25 m, although larger depths have been sampled. The draw back of the method may be settlement and separation of particles when the water flow is inadequate. Because of the large length of tubing that may be required, contamination may become a problem. Polythene or teflon is very useful. Flush the tubing with at least two times its content with water from the sample location before collection. Because of the pressure already building up by the pump, on line filtration is a possibility. Beware of contamination when using in line filters, because of their relatively large surface to volume ratio. The tubing may be lowered along a hydrowire or polythene line until the inlet is at the desired depth. Integrated sampling is easy. Strong currents may give problems in the sampling operation.

*Method #3. Bucket.* This type of sampling is apt to contamination, even though the bucket is cleaned before operation. Also, as sub-samples (bottles) are filled from the bucket, settlement of the particulate material may occur. The bucket can only collect surface samples, the surface microlayer cannot be avoided.

**Remarks:**
Certain materials from which samplers are made may not be appropriate for specific analyses. Check with the analytical procedures. Neither a free filling bottle or a pumping system is suitable for the collection of samples intended for the analysis of gasses. For sampling for dissolved oxygen and hydrogen sulphide see A-7.5 or A-7.13 of the analytical section.

**References:**
Grasshoff (1976); Venrick (1978); Aminot & Chaussepied (1983); Brockmann & Hentzschel (1983); Grasshoff *et al.* (1983); Leatherland (1985); Hillebrand & Nolting (1987); Berman & Yeats (1987); Gomez-Parra *et al.* (1987); Keith *et al.* (1991)

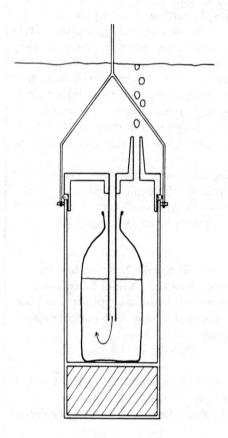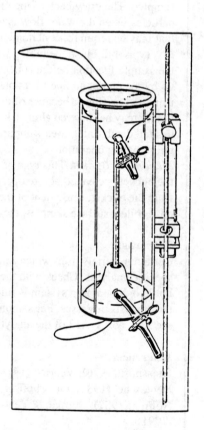

Figure 3.
*Examples of water samplers. A simple water bottle in a holder for shallow depths (l) and the Van Dorn bottle (r) (from: Head, 1985b and Van Dorn, 1962 respectively)*

**Compartment:  seston**

**Variable(s):**        **physico-chemical characteristics,**
                        **trace metals, PAHs, PCBs**

**Introduction:**
Especially in (shallow) estuaries, affected by relatively high tidal currents, the
inhomogeneity of the amount of suspended particulate matter (SPM or seston)
will usually be large. Here special attention should be given to the sampling, as
discussed in chapter 4. Possible techniques involve: sampling directly in the
sample bottle, or using a bucket, water samplers, or pumping system. Both spot
samples and integrated samples can be obtained (figure 3).
How much water should be collected for the collection of SPM, will strongly
depend on the amount of seston (and the particle size). In most estuaries
0.3 - 1 litre will be sufficient for a filter of 50 mm diameter. To remove the salt
from the SPM on filters, a standardised and validated washing procedure with
distilled water should be applied. For a high yield of seston, continuous
centrifuge techniques may be applied, which enable the collection of seston from
1 m$^3$ and more. As this technique uses the density of particles as the driving
mechanism (and not the arbitrarily chosen 0.45 µm pore size), an underestimation of the lighter fractions may occur (Salim & Cooksey, 1981). The
mechanism of separation is, however, totally different from the filtration process
and discrepancies may be expected (trace metals: Duinker *et al.*, 1979; Yeats &
Brügmann, 1990; trace organics: Hermans *et al.*, 1992).
The methods of sampling relate to a large extent to those of water sampling (ad
S-6.1) and to the sampling of phytoplankton (S-6.5). The use of a sample bottle
or a pumping system may lead to settlement of the SPM, and thus to incorrect
results.
Special treatment and equipment is needed for the collection of suspended
particulate matter for the analysis of trace compounds, both organic and
inorganic. Several types of filters are in use, predominantly because different
goals require specific filter materials. Unfortunately this results in different
'typical' pore sizes for filters, *e.g.* membrane filters (cellulose acetate or cellulose
nitrate, 0.45 µm or 0.2 µm), Nuclepore type filters (polycarbonate, 0.4 µm) or
glass fibre filters (*e.g.* GF/C, 1.2 µm; GF/F 0.8 µm).
Cleaning of filters is essential and needs to be performed prior to the filtration
step. Depending on the parameter that needs to be analysed, rigorous methods
need to be applied, such as soaking in acids (trace metals 1 M HCl overnight),
organic solvents or combustion (trace organics including pollutants). In many
cases washing of the filter with sample prior to collection (discarding of the first
50 or 100 ml) is useful.
*Trace metals*. Suspended matter for trace metal analysis is collected by either
filtration or (sometimes) centrifugation. Only limited information exists on

filtration procedures. An intercalibration exercise was carried out for the collection of SPM from coastal waters (Bewers *et al.*, 1985), while *e.g.* Laxen & Chandler (1982) and Horowitz *et al.* (1992) compared different filter types for their use in trace metals analysis.

Contamination control is of the utmost importance. The filtration step is preferred in a contamination free atmosphere (clean room or at least clean bench).

*Trace organics: PAHs, PCBs, etc.* Suspended matter for trace organics analysis is collected by either filtration or (sometimes) centrifugation. For filtration one uses normally glass fibre filters with a pore size of 1.2 or 0.8 μm. The diameter can be large (e.g. 145 mm) since often large amounts of water need to be filtered to collect sufficient material for analysis. Hermans *et al.* (1992) compared different filter types for their use in trace organics analysis, and found large differences between different methodologies. There is an ongoing discussion whether 'dissolved' trace organics should be determined as a separate compartment (in addition to the particulate fraction). It has been shown that the separation is combersome and that the results depend strongly on the methods used (Hermans *et al.*, 1992). Total analysis of the compartment water seems at present one of the best options.

**Sampling methods:**

*Method #1a. Water-sampler.* The water-sampler is much preferred as it can be operated much more reproducibly and at any depth desired. Because of the (rapid) settling of seston, it is recommended to use a small volume sampler that is emptied completely (thus collecting all seston in the sampler). If a larger sampler is used, collection of the water from the sampler should be under regular shaking of the sampler to keep the seston in suspension. In highly turbid waters the best option for seston sampling is the use of a horizontally operated sampler, constructed like a tube with constant diameter, and which is directed into the current with a vane.

*Method #1b. Sample bottle.* It is simplest to lower the bottle by hand below the surface of the water and let it fill.

For sub-surface samples, a number of constructions have been designed, which either lower the bottle in the open position, or open it at the desired depth (see *e.g.* Brockmann & Hentzschel, 1983). The system may be of the displacement type, that allows a steady flow into the bottle. An active opening procedure is preferred. The bottle(s) are lowered attached to a graduated polythene line in a suitable weighted enclosure and opened at the desired depth.

*Method #2. Pump.* As for sampling water, pumping for the collection of seston has become popular for use when a ship is under way, or for large volumes of water. Several types of pumping system are used, both *in situ*, peristaltic, bellow, etc. pumps or using a vacuum behind the collection bottle. A disadvan-tage of the method may be the settlement of particles when the water flow is inadequate. Because of the pressure already built up by the pump, on-line filtration is a possibility.

*Method #3. Bucket.* As sub-samples (bottles) are filled from the bucket, settlement of the particulate material, even under continuous movement of the bucket while sub-sampling, may occur. The bucket can only collect surface samples. This method is therefore not recommended.

**Remarks:**

**References:**
Eleftheriou & Holme (1984); Fleeger *et al.* (1988); Yeats & Brügmann (1990); Keith *et al.* (1991); Hermans *et al.*, (1992); CCME (1993)

**S-6.3**

---

**Compartment:  sediment**

**Variable(s):       physico-chemical characteristics,**
**                         trace metals, PCBs, PAHs**

**Introduction:**
Sediment is usually heterogeneous in nature. The collection of one sample for a physical-chemical characterization will almost certainly not be characteristic for the area. Multiple sampling and analysis offers better possibilities. Separate samples may be pooled (homogenized) if necessary, but this will destroy any information on the structural differences. Often samples will be collected in conjunction with the collection of benthic biota, where the combination of samples is out of the question. The interrelation between sediment characteristics and biological results is considered valuable. A review of various sampling devices can be found in *e.g.* Fleeger *et al*. (1988), Eleftheriou & Holme (1984), Baudo (1990) and Mudroch & MacKnight (1991). For an extensive bibliography one is referred to Elliott *et al.* (1993). For sediment analysis undisturbed samples are more easy to interpret. Cores will provide such samples, the actual depth of sampling (sediment thickness) can be measured, and even sub-sampled. Grab samplers collect disturbed samples, with an undefined, but rather limited sediment depth.
The use of suction or airlift samplers, applied from a ship or by diver to collect organisms is not common practice in estuarine research. As these methods try to eliminate the sediment already *in situ*, it will be obvious that the sediment can not be sampled.
Sediments tend to be inhomogeneous because of the action of (bio)turbation. Organisms (and current or wave action) may 'sort' particles according to size and density, which will also result in sorting different organic and inorganic constituents.
To minimize the non-representativity of the sample, a number of sub-samples (5-10) may be collected from one large sample (boxcorer) or independently; they are pooled into one composite sample which is then thoroughly homogenized.
When different sediment layers are to be sampled it will depend on the purpose of the sampling programme what depth(s) are to be collected. For the collection of organisms one tends to collect thicker layers (several cm) than for the geochemical and pollution related properties, where sometimes subsequent slices as thin as 1 mm are collected.
For the sampling of pollutants in sediments it will be dependent on the aim of the programme whether bulk samples, only the top layer or series of consecutive slices (pollution history) need to be sampled. For pollution studies the need for contamination control is imperative.
Normalization is used as a procedure to compensate for the effects of natural differences in sediment composition on the measured variability of pollutant

concentrations in sediments (grain size effects). Several methods have been developed to normalize for contaminants in sediments, but in practice the variety of approaches has led to a great diversity; harmonization of sediment monitoring methods has not yet been achieved (see *e.g.* by Salomons & Förstner, 1984). The fraction <20 μm (*e.g.* Ackermann *et al.*, 1983) has been adopted, which usually contains higher concentrations of pollutants than the coarser fractions (Klamer *et al.*, 1990). The fraction <63 μm (silt + clay) has become a frequently used standard 'fine' fraction. Förstner & Salomons (1980) stress its importance and argue that (1) the majority of the pollutants are contained in this fraction, (2) the (re)suspended particulate matter falls in this size range, (3) it is easily obtained by sieving (as is the <20 μm fraction) and (4) there are many data sets available enabling a better comparison.

*Trace metals.* Most sampling devices can be used for trace metal studies, provided that they are made of non-contaminationg materials (no rusty Van Veen grab samplers). Core catchers should be removed or replaced by a plastic version. Stainless steel is suitable only when the analyses do not involve chromium or nickel. PVC or perspex corers are very useful. When one feares for contamination from the corer, the middle part of the sediment core can be separated. A typical (minimum) amount needed for trace elemental analysis a few to tens of grammes.

Total sediment is often analysed and then afterwards normalized for grain size effects using Al, Sc, or Li (Loring, 1988; 1991) and Förstner (1989). Another approach involves the separation of a fine fraction by sieving (*e.g.* the fraction <63 μm or <20 μm) which is then analysed. Sieves need to be of nylon to avoid contamination.

*Trace organics: PAHs, PCBs, etc.* Stainless steel or glass devices are much preferred, but not always practical. When the construction does not allow for the use of these materials, care should be taken not to collect sediment that has been in contact with plastic surfaces.

Total sediment is often analysed and then afterwards normalized for grain size effects using total organic carbon (POC). The problem here is that in whole sandy sediments the concentrations of pollutants and of organic carbon are often close to or even below the detection limit. The analysis of fine fractions usually avoids this problem (DiToro *et al.*, 1991; Koopmann *et al.*, 1993). This approach involves the separation of a fine fraction by sieving (*e.g.* < 63 μm) which is then analysed (and which may even be normalized using POC). Sieves need to be of stainless steel.

**Sampling methods:**
*Method #1a. Handcorers.*
*Method #1b. Gravity corers.*
*Method #1c. Piston corers.*
*Method #1e. Vibro corers.*
Several types of corers are available: gravity-corers, piston-corers, vibro-corers. In shallow waters and in the intertidal zone a simple perspex or PVC tube will be a very useful corer. For sediment analysis the size is not really important. If

multiple analyses are to be performed sufficient sample material has to be collected, thus a tube diameter of approximately 8 cm is recommended. When composite samples are collected, about 10 samples (2 cm Ø corers, *e.g.* cut large syringes, see S-6.8.1) of defined length are mixed (homogenised) in a wide-mouth polythene bottle. Collect the top 2 cm of the sediment. This is not the sediment depth that is collected for the various organisms (either 5 or 25 cm, see S-6.8, S-6.10, S-6.11), but this sedi-ment layer is considered representative for the analysis of the physico-chemical characteristics.

*Method #1d. Box-corer.* Box-corers are only useful in large estuaries, where a sufficiently large ship can manoeuvre. In addition their use may be limited to low (tidal) current situations under estuarine conditions. Box-corers are specialised equipment, of considerable size and weight. Once they are retrieved, sub-samples can be collected by sub-coring using plastic tubes. The major advantage of box-corers is that they collect undisturbed samples, and contain sufficient material for all types of analysis, both of the sediment and biota. The results will therefore be better linked than when separate corers are applied. Collect the top 2 cm for physico-chemical characteristics.

*Method #2. Grab-sampler.* Various types of grab-samplers are currently in use, *e.g.* several versions of the Van Veen Grab, the Petersen-, Smith-McIntyre- and Day-samplers. Riddle (1989) compared the sampling efficiency of these samplers, including their 'bite' profiles and depth of sampling. For sediment collection an undisturbed sample is much preferred, together with a reasonable sampling depth (at least 5 cm). The chain-rigged Van Veen and the Petersen grabs are not ideal for this purpose.

Most grab-samplers are self-operating: when lowered to the sediment, they will close automatically when lifted. It is likely that not all samples will have the desired quality, because *e.g.* a stone is trapped between the jaws, the sampler didn't penetrate far enough into the sediment, etc. Collection of a new sample is strongly advised when this occurs.

Samples of the sediment should be collected in such a way that the vertical structure is preserved as much as possible. A small plastic corer can prove a useful tool. Contamination of the sample should be prevented.

**Remarks:**

**References:**
Eleftheriou & Holme (1984); Fleeger *et al.* (1988); Keith *et al.* (1991); Mudroch & Bourbonniere (1991); Burton (1992); CCME (1993)

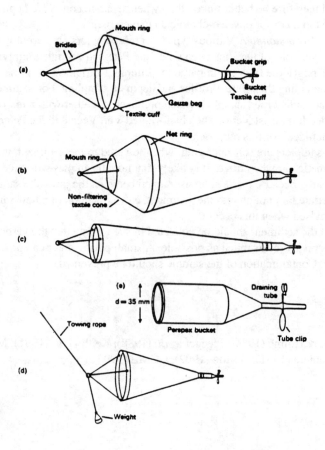

*Figure 4.*
*Examples of various types of plankton nets (From: Sournia, 1978)*

**S-6.4**

**Compartment:   water**

**Variable(s):        bacteria**

**Introduction:**
Two major approaches are followed when sampling for bacteria: filling the
(sterilized) sample bottle directly with the sample, using a pumping system or
using a sterilized water sampler. In both cases precautions should be taken to
prevent contamination of the sample. For simple counting work, extreme
precautions (sterilizing all equipment) are not essential, but the sample bottles
should still be sterilized.
Sample bottles for the collection of bacteria should not be prewashed with water
from the field. To overcome patchines in the field, especially in case aggregates
- with adhering bacteria - are present, it is recommended to sample a number of
times ($\approx 20$), and pool the sample before analysis.

**Sampling methods:**
*Method #1. Sample bottle.* Direct sampling in the bottle can provide an effective, reliable and cheap method of collection. When the sample is taken by hand
only surface waters may be collected, using a mounting device as discussed
under S-6.1.1b, only the version that is lowered in the closed condition can be
used. The maximum depth of sampling is then limited, but this is usually not a
problem in estuarine waters. Care should be taken not to contaminate the sample
with material that comes from the hands or gear. Preferably open the bottle
under water. For sub-surface sampling a set-up has been described by
Brockmann & Hentzschel (1983).
*Method #2. Pumping system.* Lowering the tubing of a pumping system to the
desired depth provides a useful means of collection. To prevent contamination
from previous samples, the tubing and pump should be cleaned by flushing with
water from the sampling location for at least three minutes.
The sample is collected in sterilized sample bottles which have not been rinsed
with water from the field. The sample bottle should be filled in a clean area,
preferably in a laminar flow clean bench with UV lighting.
*Method #3. Sterile water sampler.* Special water samplers that can be sterilized
have been developed for the collection of micro organisms (Grasshoff, 1976).
They are considered specialist equipment and the user is referred to their
instruction manual. They are all of the type that they can be opened at the
desired water depth. General remarks on the use of water samplers have been
given under S-6.1. The sampler should be emptied in a clean area, preferably in
a laminar flow clean bench with UV lighting.

**Remarks:**

**References:**
Grasshoff (1976)

**S-6.5**

---

**Compartment:  water**

**Variable(s):       phytoplankton**

**Introduction:**
The collection methods for the study of phytoplankton will to some extent
affect the diversity of the species. Water depth (spot samples) or water column
collection  (integrated samples) will almost always give different results. The
use of a water sampler or the direct filling of a sample bottle seems the simplest
method to standardize for estuarine work, as it is commonly used and a technique that will be available in all laboratories. Pump systems and nets are useful for large quantities of material, as they offer possibilities for concentration
of the phytoplankton. Water samplers, sample bottles and pumping systems
may collect spot samples, while pump systems and nets may give integrated
results. The use of Plankton Recorder systems (*e.g.* Robinson & Hiby, 1978) is
considered beyond the scope of routine estuarine sampling programme.
One should be aware of the usually large patchiness of plankton in the water
column (Duston & Pinckney, 1989).

**Sampling methods:**
*Method #1a. Sample bottle.* Collection of water samples direct into a sample
bottle, or using these bottles as a sampler and transferring the sample later, is
an adequate method. The most simple method is to lower the bottle by hand
below the surface of the water and let it fill. Sampling of the surface microlayer
should be avoided.
For sub-surface samples, constructions have been designed which either lower
the bottle in open position, or open it at the desired depth. The latter may be of
the displacement type, that allows a steady flow into the bottle. In estuarine
waters the depth limitation of the method is usually not a problem.
Lowering in the open position collects samples not at discrete depths but some
sort of integrated sample. However, collection is not regular over the sampling
path, because of the pressure difference. An active opening procedure is therefore
preferred. The bottle(s) are lowered attached to a graduated polythene line in a
suitable weighted enclosure and opened at the desired depth.
*Method #1b. Water sampler.* Methods for the collection of samples for dissolved compounds or for seston, including phytoplankton, are the same. Any
water sampler can be used. However, sampling is best performed using relatively small water samplers (1-2 litres), of which the total water content can be
used. As part of the phytoplankton may have a tendency to be deposited, an
overestimation of some species may occur if only part of the sampler content is
collected. If this is not possible, the sample in sampler should be homogenised
(reversing the sampler) before sub-sampling.
General remarks on the use of water samplers are given in S-6.1. The sampler

is mounted on a weighted hydrowire, and closed at the desired depth by a dropping weight. In case of strong currents when the ship is at anchor, the sampler may be used sliding along a separate hydrowire that is kept almost vertical using a type of 'depressor' as used in zooplankton catches in open sea. In all cases the actual depth should be estimated as accurately as possible using a graduated line. Assume the middle of the sampler is the actual sampling depth and correct for deviation from the vertical.

*Method #2. Pumping system.* Several types of pumping system are used, both peristaltic, bellow, membrane, etc. pumps or using a vacuum behind the collection bottle. Centrifugal pumps are not to be recommended because they may be highly damaging to the plankton. Most pumps will provide sufficient pressure, provided they don't have to suck the water to a considerable height first (static suction head). The practical sampling depth is to about 25 m, although larger depths have been sampled. A disadvantage of the method may be the settlement of particles when the water flow is inadequate. Also, fragile species might be physically damaged. Flush the tubing with at least two times its volume of water from the field site before collection. In between operations the tubing may get coated on the inside with organisms, which can be a serious source of contamination. A larger diameter tube will have less total frictional loss than thin tubing given a constant discharge. The tubing may be lowered along a hydrowire or polythene line until the inlet is at the desired depth. Strong currents may give problems in the sampling operation. The volume that has been filtered should be recorded and for statistical reasons should not vary between sampling events. A standard volume of 50 litres is proposed for estuarine work. Both spot sampling and integrated sampling are easy to perform.

*Method #3. Nets.* The major advantage of nets is the relative ease of filtering large volumes of water and the resulting concentration of species (figure 4). The use in small estuaries may be limited, however. A major drawback of the method is the selectivity of the nets for species size and form, which is dependent on the mesh-size and the clogging of the net during collection.

Because of this, information on the net gauze and the volume of water filtered has to be presented together with the results.

For phytoplankton studies in estuaries standard small nets (mouth diameter 15 cm, length 110 cm; Tangen, 1978) with a gauze of 10 µm are proposed.

Vertical hauls collect material from the entire water column. The weighted net is lowered from the drifting ship to the bottom, and withdrawn slowly. When this is performed from a moving ship (oblique haul) the total amount of material collected is larger, but in small estuaries the chance of mixing different phytoplankton assemblages, because of changing physico-chemical conditions, is evident. Towing speed should not exceed 1 m/s (ca 2 knots). In estuarine waters clogging of the nets can be a major problem.

**Remarks:**

**References:**
Beers (1978); Sournia (1978); Throndsen (1978)

**Compartment:   water**

**Variable(s):        zooplankton**

**Introduction:**
The large diversity in groups of organisms, their size and life-stage make it
difficult to establish one generally accepted method for sampling. Again several
methods are given that are not very comparable, but which are commonly used
and require equipment which is not too complicated. Independent of the
collection method a mesh-size has to be defined. Proposed here as a preferred
standard method for zooplankton sampling is a mesh of 55 or 63 µm.
The Methods #1, #4 and #5 sample 100 litres water which is poured through a
conical net as a standard procedure, while the other methods require large
volumes of water, the organisms being collected directly in the nets. A volume
of 5 m$^3$ should be the preferred volume. For estuarine work these methods are
considered less practical.
The nets are all sprayed from the outside with water from the field, to allow the
zooplankton on the inside of the net to be collected in the funnel at the bottom.
A wash bottle is a convenient tool here.
One sample per location is a minimum requirement, which should be collected
at mid water depth.

**Sampling methods:**
*Method #1. Pump & net.* Several types of pumping system are used, especially
bellow, membrane pumps. Centrifugal pumps are not recommended because
they may be highly damaging to the zooplankton. Actual speed of pumping is
very important, to avoid missing fast swimming plankton. One drawback of the
method is, that it is not known to what extent more active organisms can avoid
the mouth of the pump. Also, fragile species or life stages might be physically
damaged. A larger diameter tube will have less total frictional loss than thin
tubing given a constant discharge. The tubing may be lowered along a hydrowire or polythene line until the inlet is at the desired depth. Strong currents
may give problems in the sampling operation. The volume that has been filtered should be recorded and for statistical reasons should not vary between
sampling events. A standard volume of 50 litres is proposed for estuarine work.
Both spot sampling and integrated sampling by moving the tube entrance
vertically through the watercolumn, is easy to perform.
*Method #2. Towing nets.* For large amounts of zooplankton, or very low
densities, this procedure has advantages. In estuaries there is probably no need
for this procedure. A drawback of this method is the difficulty of estimating the
volume of water passed through the net. Clogging may, depending on the mesh-
size, be a problem in estuaries. Sampling is performed from a moving ship; in
small estuaries there is a strong chance of mixing different zooplankton

assemblages, because of changing physico-chemical conditions. Towing speed is critical for the quantitative collection of the plankton (figure 4).

*Method #3. High-speed samplers.* High-speed samplers are not very useful for estuarine work for the reasons mentioned in Method #2.

*Method #4. Collection tube & net.* For the integrated collection of a vertical cross section of the upper ca 4 metres of the water column a tube sampler has been used. The design involves a PVC tube of about 12 cm diameter. At one end a ball valve is mounted. When the tube, with the valve at the base, is quickly inserted vertically into the water, a water-column can be collected. The contents are then poured through the net. By repeating this procedure the de-sired 50 litres can be filtered. The handling of a sizeable length of the tube is limited to about 4 - 5 m, but this may be sufficient for estuarine work. Fast moving organisms probably escape from the tube mouth, but will be collected more quantitatively than by using a pumping system.

*Method #5. Bucket & net.* For the collection of surface water zooplankton the use of a bucket which is emptied in the conical net offers a simple and effective means of collection. A 10 litres graduated bucket thus requires 5 samples to be filtered.

**Remarks:**

**References:**
Tranter & Fraser (1968); Raymont (1983); Tett (1987)

**Compartment: water**

**Variable(s): hyperbenthos**

**Introduction:**
The hyperbenthos is defined as the mobile fauna living in the lower part of the water column but more or less dependent on the proximity of the bottom. It can be considered as the uppermost part of the benthic community. Numerous animals belonging to a variety of taxonomic groups occupy the hyperbenthal. The permanent residents are predominately mysids and other crustaceans (mainly amphipods and isopods). Other animals spend only part of their life cycle in the hyperbenthal (*e.g.* larval crabs, shrimps and fish). Most hyperbenthic animals are fast swimmers. Since the study of the hyperbenthic fauna is relatively new in marine studies, there is as yet no standardization in definitions or sampling gear. The hyperbenthos goes under a variety of names (*e.g.* suprabenthos, nectobenthos, demersal zooplankton, bottom-plankton, near-bottom fauna) and is caught with a plethora of sampling devices.
Sampling of the hyperbenthos recquires specific equipment which, in comparison to devices used to sample other benthic or planktonic compartments of the estuarine ecosystem, are quite complicated and expensive. Usually sledges are deployed which are trawled with a ship. Alternatively, passive fishing techniques can be used. These can be useful for sampling in areas too shallow to be sampled by ship or in zones where no stretches of sufficient length can be found which are free from obstacles. Diurnal and tidal vertical migrations of hyperbenthic animals can be studied with high-speed samplers.
A sledge type gear equipped with 0.5 mm or 1 mm mesh nets is proposed here as a preferred standard method, when used over defined tracks and times.

**Sampling methods:**
*Method #1. Sledges.* Several types of sledges have been used. Usually they are rather heavy constructions consisting of a metal frame with one or several nets. Mesh size should be 0.5 or 1 mm and the lower 1 metre of the watercolumn should be sampled, though it is often impossible to sample the lower 10 cm. A design with two or more nets on top of each other is preferred. The sledge should be equipped with an automatic opening-closing device, preventing contamination with organisms from the upper water layers. The distance sampled, defined as that stretch where the sledge is in contact with the bottom, and the volume of water filtered through the net should be measured. Trawling can either be done with the tide (standard of 500 to 1000 metres at a ship speed of 4.5 knots relative to the bottom) or against the tide (standard of 5 minutes at 1.5 knots). Sampling should be carried out during daytime, when hyperbenthic animals are known to concentrate near the bottom. Due to the weight of the gear, a sledge must be operated from a relatively large research vessel, limiting its use for shallow sampling stations. Nets should be sprayed with water from the field to allow the

hyperbenthos inside the net to be collected in the funnel at the bottom. One sample per location will usually suffice in estuaries.

*Method #2. Passive fishing techniques.* Plankton nets or fyke nets (mesh size 0.5 mm or 1 mm) can be placed with their mouth facing the current. A fishing time of 5 to 10 minutes will usually collect sufficient material in estuaries. Samples can only be taken when current velocities are relatively high in order to minimize net avoidance.

*Method #3. High-speed samplers.* High speed samplers (mesh 0.5 or 1 mm) do not efficiently catch the most active swimmers of the hyperbenthos. This method is only recommended for sampling hyperbenthic animals at different depths to assess vertical migrations patterns.

**Remarks:**

**References:**
Brunel *et al.* (1978); Bhaud (1979); Dauvin & Lorgère (1989); Hamerlynck & Mees (1991); Mauchline (1980); Sorbe (1983).

**Compartment: sediment**

**Variable(s):        micro-phytobenthos**

**Introduction:**
Micro-phytobenthos significantly contributes to the primary production of
estuarine areas and is present not only at the very surface of sediments, but also
in deeper layers (Cadée, 1974; De Jonge, 1992). Therefore the top 0.5 cm
should be collected for mass production and the top 2 cm for determining
biomass. This 2 cm should be subdivided into 0 - 0.5 and 0.5 - 2 cm instead of
the very surface layer.
Composite samples should be collected to avoid the influence of supposed
heterogeneous distribution of the micro-flora. A minimum of 20 individual
samples should be collected along one (or more) pre-described transects at 1 m
distance intervals. For optimal work 50 to 100 samples should be collected and
pooled.
Due to the turbidity of most estuarine waters, the micro-phytobenthos will be
most important in the intertidal areas.

**Sampling methods:**
*Method #1. Hand corer, 2.5 cm Ø.* Hand operated corers are almost exclusively used. A plastic tube of the proper diameter is introduced into the sediment. A sawn-off syringe is very useful, as it also provides a plunger that keeps
the sediment in place when the tube is withdrawn. The excess sediment is easily
removed by pressing the plunger, so the desired length of sediment (0.5 or 2 cm)
is retained.
*Method #2. Box corer, 2.5 cm Ø (sub-samples).* For subtidal areas box core
samples can be used, where sub-samples are collected using the syringe
described under Method #1. A minimum of 5 box cores should be collected,
and from each 4 replicate samples can be taken and homogenized into a
composite sample.

**Remarks:**

**References:**
Cadée & Hegeman (1974; 1977); De Jonge (1992)

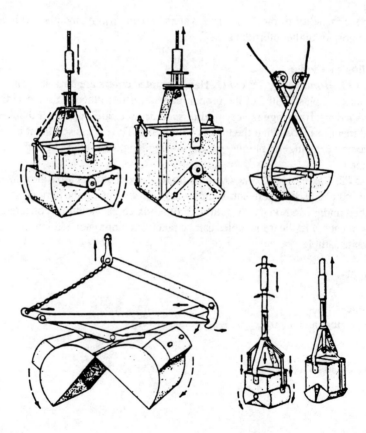

*Figure 5.*
*Examples of grab samplers (from: Fleeger et al., 1988)*

**Compartment:   sediment**

**Variable(s):       macro-phytobenthos**


**Introduction:**
Macro-phytobenthos is not of importance in all estuaries. Dominant species,
those that exhibit a relatively large portion of the biomass, will, however, be of
interest. As stated in the preface, salt marsh species are not considered here, and
the study is limited to intertidal macrophytes. Most important in this respect are
the macro-algae *(Fucus* spp., *Ulva* spp., *Enteromorpha* spp.) and seagrasses
(*Zostera* spp.). Apart from specificity for certain salinity regimes, macrophytes
are rather sensitive to bathymetric gradients. Therefore species abundance and
biomass should be related to this variable.
In areas where there is a large variation in sediment type or (bathymetric)
gradient, only those areas should be sampled that cover a relatively large
proportion of the intertidal area in the salinity regime that has to be investigated.
Sampling of non-representative areas should be avoided.
Sfriso *et al*. (1991) studied the effect of different sampling strategies in the
shallow waters of the Venice Lagoon. They tested the effect (on average
biomass) of the number of sub-samples that should be collected. Assuming that
20 sub-samples would represent a good representation of the area studied, they
found that usually 4-8 subsamples were to be considered representative. Of the
two sampling methods given, the first is usually applied in intertidal areas. The
second on the other hand, serves well in subtidal substrates.
Aquatic macrophytes accumulate significant amounts of trace metals from
ambient waters and the concentrations of elements present in tissues of these
plants can proveide information on the extent of contamination of estuarine and
coastal waters (Phillips, 1994). Especially species of brown and green algae, but
also seagrasses have thus been used. In contrast to the many trace metal studies,
macrophytes have only rarely been used to quantify the abundance of trace
organic contaminants, which is probably due to the low lipid content of these
organisms.
For pollution studies it is very important that the surfaces of the plants are
washed carefully, as both organic (*e.g.* epiphytes) and inorganic (adhered
particulate matter) contamination will otherwise occur.
For trace metal studies separate samples are to be collected, taking care not to
contaminate the tissue. Transport is performed in plastic polythene bags in a cool
environment. Samples may be stored deepfrozen until chemical analysis.

**Sampling methods:**
*Method #1. Selected 10 x 10 cm samples.* A minimum of four 1 m$^2$ quadrats
should be selected in the area of investigation, on a transect perpendicular to
the bathymetric gradient. These squares are to be selected randomly, but must

be representative of the intertidal macro-phytobenthos in the salinity regime area. Within these square metres four sub-areas of 10 x 10 cm are randomly selected (using randomization tables, making 16 samples all together) and all macrophytes that are contained in this surface area (at the base of the plant) are collected by cutting. As the roots (if available) are difficult to collect quantitatively, they are not collected. Other surface areas may be desirable, when thin densities are observed.

*Method #2. Selected 100 x 100 cm sampling*. At representative locations an area of 15 x 15 m is selected. Within these areas a minimum of four (but eight is preferred) randomly selected 1 m$^2$ squares are selected. An aluminium box of 1 x 1 x 0.7 m is efficient for this purpose. All macro-algal material is collected within this box by cutting. As the roots (if available) are difficult to collect quantitatively, they are not collected.

**Remarks:**

**References:**
Polderman (1978); Pringle (1984); Sfriso *et al.* (1991); Phillips (1994)

**Compartment:  sediment**

**Variable(s):       meiofauna**

**Introduction:**
Where the meiobenthos is concerned there is a large variation between various locations and different types of sediment. For practical reasons it is not adequate to prescribe one single method of collection. There is no optimum between surface area (corer diameter) and number of organisms per sample (underestimation vs counting problem). In practice, experienced scientists have arrived at a procedure that involves somewhat larger size samples in sandy sediment, and smaller in muddy sediment. Others prefer one standard size (*e.g.* 3.5 cm) and depth (10 cm) and take replicates in case of low densities of organisms. As the ultimate goal is (among others) an estimation of the abundance of the different species, the different core sizes for the different sediments is not necessarily a problem. Therefore two corer sizes are recommended here.

For subtidal work the use of a box-corer is much preferred, from which subsamples of the appropriate diameter can be collected. The major problem of box-corers is their availability and size. Various gravity or piston cores can be used instead. Grab-samplers will often be used for subtidal sediment sampling because of availability, but a rather disturbed sample will result, and much of the fine surface material will be washed away by the bow-wave of the grab, loosing a high proportion of certain taxa, especially copepods.

Intertidal areas are easier to reach and to sample. Hand operated corers are then preferred.

A minimum of four individual replicate samples should be collected on a transect, at about 10 m intervals.

Standard sampling depth is 5 cm. In case deeper samples are required, 15 cm becomes the second standard (especially in clean sand where the meiofauna penetrate deeper).

If the density of the organisms is expected to be the too high for the given tube diameter, the sample should be split once or twice over the length. Take care to apply a separation parallel to the sides of the core, weigh the subsamples for corrections in surface area. This method may be called inaccurate, and it is preferable to split the homogenised preserved sediment sample with a special sample splitter (Pfannkuche and Theil, 1988).

(Sediment) samples should be stored after addition of 4 % formalin in warm (60 °C) seawater solution, in polythene bottles prior to analysis. A warm solution of formalin is advised, to prevent nematodes rolling up, which will make identification nearly impossible.

Because of the different logistical problems encountered in sampling intertidal and subtidal regions, different approaches are given here.

**Sampling methods (subtidal):**

*Method #1a. Box-corer, 2 cm Ø sub-samples (mud).* Box-corers are specialised equipment, of considerable size and weight, offering optimal possibilities
for collecting undisturbed sediment samples (see figure 8, p. 218). Once they are
retrieved, subsamples can be collected by corer using plastic tubes of the desired
diameter. Another advantage of this method of collection is that the cores contain
suf-ficient material for all types of analysis. The results will therefore be better
linked (*e.g.* to the physico-chemical characterization of the sediment) than when
separate corers are applied.

*Method #1b. Box-corer, 8 cm Ø sub-samples (sand).* This method only differs
from the previous one by the diameter of the sub-sample collection tube. Instead
of withdrawing the corer, it may be dug out.

*Method #2a. Hand operated corer, divers, 2 cm Ø (mud).* Divers are able to
position (small) coring tubes with care and can push the corer slowly into the
sediment. This has the added advantage that the exact sampling spot can be
visually inspected. A sawn-off syringe is very useful, as it also provides a
plunger that keeps the sediment in place when the tube is withdrawn. For longer
corers plastic tubes serve well.

*Method #2b. Hand operated corer, divers, 8 cm Ø (sand).* This method is
identical to method #2a except for the diameter of the coring tube.

*Method #3. Bow wave free corers.* A special type of subtidal corer has been
developed that eliminates the shock or bow wave that develops in front of the
free falling or pole corers (see Method #4). Good design features include slow
sediment penetration, large flow-through tubes and a trip mechanism that does
not interfere with the water flow through the corer tube or disturb the sediment
before penetration (Fleeger *et al.*, 1988). Examples that are useful in estuarine
waters include the Craib-sampler.

*Method #4. Corers.* For water depths up to about 4 m pole-samplers can be
used. The consist of corers attached to a pole and can be operated manually
from a boat. For deeper subtidal collection of sediment, remote samplers like
piston- or gravity-corers can be used. They all suffer, however, from the shock
wave that builds up in front of the opening of the corer, and which may blow
away the meiobenthos that is present at the very surface of the sediment. This
will result in an underestimation of (surface dwelling) meiofauna species.

*Method #5. Grab sampler.* Grab samples differ very much from the previous
methods. Various types of grab-samplers are currently in use, *e.g.* several
versions (figure 5) of the Van Veen Grab, the Petersen-, Smith-McIntyre- and
Day-samplers. Riddle (1989) compared the sampling efficiency of these
samplers, including their 'bite' profiles and depth of sampling (figure 6). For
sediment collection an undisturbed sample is much preferred, together with a
reasonable sampling depth (5 cm). The chain-rigged Van Veen and the Petersen
grab do not seem suitable for the purpose. Grab samplers also suffer from the
bow wave effect. Most grab-samplers are self-trigerring: when lowered to the
sediment, they will close automatically when lifted.

It may be that not all samples will be of the desired quality, because *e.g.* a
stone is trapped between the jaws, the sampler didn't penetrate far enough into

the sediment, etc. A new sample should be collected when this occurs. Sub-samples of the sediment should be collected in such a way that the vertical structure is preserved as much as possible. A small plastic corer (sawn off syringe) can prove a useful tool.

**Sampling methods (intertidal):**
*Method #1a. Hand-corer, 2 cm Ø (mud).* Because of the easy access to inter-tidal areas, hand operated corers are almost exclusively used. A plastic tube of the appropriate diameter is pushed into the sediment. A sawn-off syringe is very useful, as it provides also a plunger that keeps the sediment in place when the tube is withdrawn. The excess sediment is easily extruded by pressing the plunger so that the desired length of sediment (5 cm) is retained.
*Method #1b. Hand-corer, 8 cm Ø (sand).* A PVC or perspex tube is inserted in the sediment. The corer should preferably have a piston, which facilitates the withdrawal of the sediment. Another solution to overcome the impossibility to retrieve the sediment core, is to insert the tube, fill the tube with local water to the top, place a rubber stopper on the tube. Extract the corer from the sediment , place a stopper at the bottom end. Then remove the top stopper, and finally siphon the water off (figure 7). Retain a sample of 5 cm length. Instead of withdrawing
the corer, the corer may be dug out.

**Remarks:**

**References:**
Craib (1965); Warwick (1984); Higgins & Thiel (1988); Fleeger *et al*. (1988); Pfannkuche & Thiel (1988)

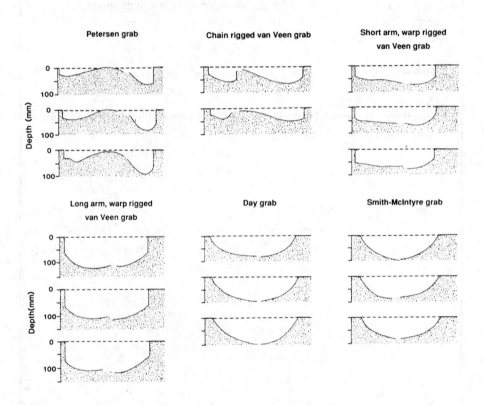

*Figure 6.*
*Differences in bite profiles of six grab samplers (from: Riddle, 1989b)*

**Compartment:  sediment**

**Variable(s):        macro-zoobenthos**

**Introduction:**
In areas where there is a large variation in sediment type, only those sediment
types that cover a relatively large proportion of the subtidal area in the salinity
regime that has to be investigated, should be sampled.
Preferably sampling should be conducted in the shallow subtidal and the
intertidal zones. In the deeper subtidal areas the abundance of the macro-
zoobenthos will be poorer, especially when the deeper parts are formed by tidal
channels. Due to high current velocities few species/numbers are able to live
there. This is not considered a typical estuarine habitat.
For subtidal work the use of a box-corer is much preferred. The major problem of
box-corers are their availability and size. Therefore, grab-samplers will be often
used for subtidal sediment sampling. For intertidal work hand corers of different
sizes are almost exclusively used. One standard size is proposed, as differences
may occur when collection is carried out with different core diameters.
The standard sampling depth is 25 cm. This can be achieved using a boxcorer
or handcorer, but may be more difficult when using a grab sampler (figure 5).
A minimum of 4 individual replicate samples should be collected either randomly
or along a pre-described transect at 10 m distance intervals, in order
to get representative information on the subtidal macro-zoobenthos. Sampling
depth will also be dependent on the type of grab sampler used (Riddle, 1989b)
(figure 6).
It has been found that tidal zonation may influence the benthic fauna distribution
and macrozoobenthos related variables. Therefore three sites; *i.e.* the upper, mid
and lower tidal level may be a necessary part of the sampling strategy. It will be
clear that the number of replicates should be applied to each of these spots, thus
multiplying the number of analyses by three.
The organisms should be extracted from the sediment using a sieve with a
1 mm mesh. In fine sediments containing mainly oligochaetes sieving over a
0.5 mm screen may be more appropriate (Bachelet, 1990).
The samples should be stored separately in polythene bags or bottles.
Add 6 % formaldehyde in seawater to the core residues for preservation and
storage. For biomass estimations, the use of glutaraldehyde is preferred,
however (5%).

**Sampling methods (subtidal):**
*Method #1. Boxcorer.* A boxcorer must be operated from a relatively large
research vessel. The samples should be collected either randomly or along a
pre-described transect at 10 m distance intervals, in order to get representative

information on the subtidal macro-zoobenthos. The minimum number of samples that should be per station/area is 5 - 10 boxcores (see figure 8, p 218). *Method #2. Grab sampler (0.2 m²).* Grab samples differ very much from the boxcoring method and the coring methods used in intertidal areas. Various types of grab-samplers are currently in use, *e.g.* several versions of the Van Veen Grab, the Petersen-, Smith-McIntyre- and Day-samplers. Riddle (1989b) compared the sampling efficiency of these samplers, including their 'bite' profiles and depth of sampling. For collection of macro-zoobenthos an undisturbed sample is much preferred, together with a reasonable sampling depth. The chain-rigged Van Veen and the Petersen grab do not seem suitable for the purpose. Most grab-samplers are self-trigering: when lowered to the sediment, they will close automatically when lifted. It may be that not all samples are of the desired quality, because *e.g.* a stone is trapped between the jaws, the sampler didn't penetrate far enough in the sediment, etc. A new sample should be collected when this occurs.

The 5 - 10 replicate samples should be collected on a transect, at about 10 m intervals.

### Sampling methods (intertidal):

*Method #1. Handcorer (200 cm²).* A hand operated corer consists of a PVC or stainless steel tube of ca. 16 cm Ø. It should be possible to close the top (reduced) opening with a rubber stopper. In some sediments the air above the sediment during collection prevents collection of the sample due to expansion. In those cases, fill the space on top of the core completely with water, close the top opening and extraction will be easier.

The corer is pushed into the sediment, taking care not to sample an area that has been disturbed by human action.

*Method #2. Handcorer (other size).* Other diameters of hand corers may be in use. One should be aware of the differences that may be the result from the deviating diameters (Elliot, 1977).

For operation the conditions are equal to Method #1.

### Remarks:

### References:

Holme (1964); Elliot (1977); Downing (1979, 1989); Morin (1985); Salonen & Sarvala (1985); Hartley & Dicks (1987); Wolff (1987); Vézina (1988); Riddle (1989a,b); Rumohr (1990); Essink & Kleef (1991)

# 7. ANALYTICAL PROCEDURES AND METHODS

In this chapter the most important physical, chemical and biological analyses that are important for a description of the estuarine environment are summarized. It is not the intention of the authors to reproduce all the details in handbooks on analytical chemical and biological methods in seawater. Neither are all the possible variables included, nor all possible methods per variable. Methods have been selected in view of optimum applicability for the estuarine environment, keeping in mind that not all participants have access to the newest techniques or sophisticated (automated) instrumentation.

Another criterion for the selection of any analytical technique was the possibility of a direct measurement of the variable. A number of variables cannot be quantified by measurements, but are derived by calculation from other variables. These calculations, *e.g.* the conversion of particulate organic carbon data to biomass, is based on conversion factors derived from the literature. If analytical data have to be stored in a data base, as is the case with the JEEP92 programme it is more sensible to store the raw data in the data base, and leave it to the users of the data base to perform the conversions. The data will thus be more easily useable if, for example, the conversion factor changes in the future.

Where several possible techniques or methods give identical results, this is indicated using suffixes: Method #1a, Method #1b. Sometimes the results of different techniques do not compare well in intercalibration exercises. In these cases the methods will have different numbers: Method #1, Method #2. Unless stated otherwise, Method #1 is preferred over Method #2, etc.

Some Method numbers are followed by an asterisk (*). The reason is, that these methods are based on calculation (using a measured variable) rather than on direct measurements. One should give some consideration to storing data in a data base; it is better to use the original data rather than the calculated results, as calculation methods may change with time. The asterisk only serves as a reminder, no quality statement is intended.

For each method the units in which the results should be expressed are given. These may be different from traditional units (as for example in the case of nutrients), but international guidelines are followed here: SI rules are used (Unesco, 1985). Though the expression of a concentration should preferably by per kg (instead of per litre, or dm$^3$), in this manual the "per litre" is used throughout for practical reasons. In table 5 an overview of the various analyses together with the symbols, units and analysis codes is presented for easy reference. Annex III lists all analytical methods considered in this manual. Each analytical procedure has the same lay-out. The meaning of the unique JEEP92 **analysis code** (A-7.xx number) for easy reference and used throughout the volume, the **variable** with symbol and the **unit** in which the results should be expressed, and the compartment are marked in the heading of each procedure.

In the **Introduction** to each method some general information is presented, together with information on other techniques where applicable.

**Sampling, Sample volume, Sample treatment** and **Storage** are only briefly mentioned, and the reader is referred to the sections on sampling procedures (chapter 6).

As samples are not always collected by the most specialized personnel, relatively detailed information is given when necessary on the field activities (sampling, sample treatment, preservation and storage).

In the **Analytical methods** section the method(s) of analysis are outlined.

In most cases the reader is referred to a detailed description of the method in the literature.

Several standard handbooks are available on seawater and estuarine analysis, which are considered essential. For detailed information the reader is referred to these handbooks and the original sources given therein, and to the **Method(s) references** given at the end of each analytical procedure.

Useful handbooks on marine analytical chemistry and on marine (and estuarine) biology include:

- Aminot, A. & M. Chaussepied, 1983. Manuel des analyses chimiques en milieu marin. Centre National pour l'Exploitation des Océans (CNEXO), Brest, pp. 395
- Grasshoff, K., M. Ehrhardt & K. Kremling (eds), 1983. Methods of seawater analysis. 2nd ed. Verlag Chemie, Weinheim
- Higgins, R.P. & H. Thiel, 1988. Introduction to the study of Meiofauna. Smithsonian Institution Press, Washington DC, pp. 488
- Morris, A.W. (ed), 1983. Practical procedures for estuarine studies. A handbook prepared by the Estuarine Ecology Group of the Institute for Marine Environmental Research. IMER, Plymouth, pp. 262

- Parsons, T.R., Y. Maita & C.M. Lalli, 1984. A manual of chemical and biological methods for seawater analysis. Pergamon, Oxford, pp. 173
- Rodier, J. (ed), 1984. L'analyse de l'eau: eaux naturelles, eaux résiduaires, eau de mer. 7th Ed., Dunod, Paris, pp. 1365
- Sournia, A. (ed), 1978. Phytoplankton manual. Unesco, Paris, 337
- Strickland, J.D.H. & T.R. Parsons, 1968. A practical handbook of seawater analysis. Fish. Res. Bd. Can. Bulletin 167, Ottawa, 311; ibid. 1972, 2nd edition.
- Unesco, 1985. The International System of Units (SI) in oceanography. Technical Papers in Marine Science, no. 45, pp. 124

The list of analytical procedures is sub-divided into different sections:

analyses in water (A-7.1 - A-7.26)

analyses in seston and sediment (A-7.27 - A-7.38)

analyses of biota:

bacteria (A-7.39-A-7.40)

phytoplankton (A-7.41 - A-7.43)

zooplankton (A-7.44 - A-7.47)

hyperbenthos (A-7.48 - A-7.50)

micro-phytobenthos (A-7.51 - A-7.53)

macro-phytobenthos (A-7.54 - A-7.56)

meiofauna (A-7.57 - A-7.58)

macro-zoobenthos (A-7.59 - A-7.64)

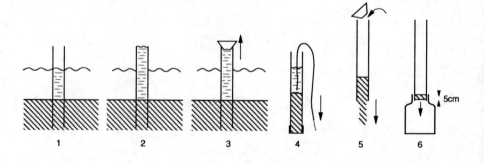

*Figure 7.*
*Sequence of operation with a hand corer on tidal flats or in shallow waters*

**Variable:**        **salinity**

                **S**

**Unit:**            $10^{-3}$ [1])

**Compartment:**    **water**

**Introduction:**
Salinity is one of the earliest defined variables in the marine environment
(Grasshoff *et al.*, 1983; Mamayev, 1975). It constitutes one of the most
fundamental variables in estuarine research, and great care should be given to
its determination and precision, and thus calibration.
Absolute Salinity is defined as the ratio of mass of dissolved material in seawater to the mass of seawater. As this quantity cannot be measured directly a
Practical Salinity is defined as a polynomial equation, which is, however, only
valid between $2 \leq S \leq 42$ (Unesco, 1985). For practical reasons we use Salinity,
which implies that there is a constant relationship between the major constituents in seawater. There are several methods of determining the salinity of sea-
and estuarine waters. The most important and practical are:
- Conductivity. As this relationship is dependent on the fixed ratio between
major seawater constituents, the method is not reliable for $S < 2 \times 10^{-3}$. This
has serious implications for use in the upper estuarine regions.
- calculation based on chlorinity, the amount of chlorine [2]). In the upper
regions of the estuary this method is more reliable than the conductivity
method, but one should beware of riverine water with relatively high chloride
content [3]). Essentially chlorinity is a separate, independent property, but for
practical reasons we may use Cl to calculate the salinity.
For less precise measurements the refractive index method may be used.
Calibration of the salinity measurement should be performed regularly by
using Standard Seawater (obtained from IAPSO Standard Sea Water Service).

**Sampling:**
- *in situ* measurement of conductivity, using a CTD or ST probe;
- collection of water directly in a bottle, using a water sampler or a pump.
The measurement of profiles is encouraged before the actual sampling starts,
to enable detection of possible haloclines. The salinity should be measured in
relation to the other samples collected, as rapid salinity changes are likely to
occur in estuarine environments [4]).

**Sample volume:**
for *in situ* measurement non relevant; others: 1 litre

**Sample treatment:**
none

**Storage:**
samples can be stored for several weeks preferably at low temperatures (but never $< 0\ °C$) in borosilicate glass bottles with screw caps. No ground glass stoppers, and no high-pressure polythene should be used.

**Analytical methods:**
*Method #1a.* Conductivity, is a measurement of a physical property of the seawater and determined either as a conductivity ratio (calibrated against standard seawater) or by direct measurement of conductivity, both of which are corrected for temperature. The conductivity-density-salinity-chlorinity relationships are defined in Unesco (1981). Usually the measurement is performed *in situ* using a Conductivity-Temperature-Depth (CTD) probe, or a more simple ST-meter; samples are collected for calibration in a salinometer.
*Method #1b.* Determination of chlorinity by titration of the chlorine with silver nitrate (Mohr-Knudsen titration), which is recalculated into salinity according to the definition: $S\ 10^{-3} = 1.80655\ Cl\ 10^{-3}$ (Unesco, 1985).
*Method #2.* Refractive index method. The refractive index of seawater is a function of salinity. An increase in salinity of $1 \times 10^{-3}$ results in an increase in the refractive index by about 0.0002. A (hand-held) refractometer is a simple instrument that can conveniently be used in the field. Although temperature effects upon the measurement are small, automatic compensation is applied. Since this optical method is hampered by suspended particulate matter, in most estuaries the sample should be filtered (*e.g.* 0.45 µm) prior to the analysis. The unknown sample is calibrated against distilled water and seawater of known salinity. The accuracy of the hand-held refractometers is 0.5-0.6 S, indicating that the method is very useful for many biological applications, where a rough indication of the water is sufficient, but not to be applied for most physical/chemical work.

**Remarks:**
[1]   According to the new definition of salinity the units are not expressed in $10^{-3}$, but as 1 (the $10^{-3}$ is then ommitted). (Unesco, 1985). Generally the $10^{-3}$ is still in use, however. Quite often ‰ is given. Not to be used is ppt as it may denote both parts per thousand and parts per trillion.
[2]   See method for chlorinity.
[3]   Millero (1984) discussed the conductivity-density-salinity-chlorinity relationships for estuarine waters. He concluded that if errors in salinity of ±0.04 are acceptable, the UNESCO definition may be applied without corrections for the ionic composition of the estuary, even in the freshwater part.
[4]   A commonly used procedure where salinity profiles are recorded and afterwards samples for various variables are collected should be avoided. If large variations in salinity occur, the salinity should be determined for

each sample taken, either during sampling (conductivity probe) or by separate analysis.

**Method(s) references:**
Martin (1968); Strickland & Parsons (1968; 1972); Aminot & Chaussepied (1983); Grasshoff *et al.* (1983); Rodier (1984); Leatherland (1985)

VII Pl.

## TWEEDE TAFEL

de Gewigten en Coleuren der zoete Put Fontein = en r = Wateren, met het Areometrum omtrent den Oever der Zee geschept.

## AREOMETRUM

Dat tot alle de waernemingen nopens het gewigt van verschei. dene zoorten van Wateren ge. bruikt is geweest, en hier in des. zelfs weezendlyke gedaente en groote verbeeld word, wegende een Once, drie Dragmen en tien Greinen. Onder aen de voet is een zoort van een kleine Bol gevult met Quik. De loode Ringen hier mede afge. schetst, en die 't gewigt hebben, dat nevens ieder gemeld word, heb. ben voor een evenwigt gedient met de vereischte hoeveelheid daer by te. voegen.

| Plaetsen | Maenden | Dagen | Coleuren | Oncen | Dragmen | Greinen |
|---|---|---|---|---|---|---|
| te Montpelier uit de Put. ten van den H.r Matt. | November | 6 | Helder | 1 | 3 | 30 |
| Montpelier de Put van S.t Giles. | November | 6 | Helder | 1 | 3 | 28 |
| Silva Royal of den Oever van de kleine Rhône. | November | 22 | Onklaer, dog in een glas stilstaen de word helder | 2 | 3 | 29¼ |
| By de Hutten van Orgon aen den Oever van de klei ne Rhône, 500 schreeden van de Zee. | November | 23 | Onklaer, dog in een glas stilstaende word helder | 1 | 3 | 30¼ |
| Putten, 3 voeten diep, door my gegraven omtrent de Hutten van Orgon, 500 schreeden van de Zee en 12 schreeden van de Rhône. | November | 25 | Onklaer, dog in een glas stilstaende word helder | 1 | 3 | 29¼ |
| By de S.t Maria's uit de putten van den Consul. | November | 26 | Witagtig | 1 | 3 | 33 |
| By de S.t Maria's uit de putten van Lombard. | November | 26 | Witagtig | 1 | 3 | 30 |
| By de S.t Maria's uit de putten van Becheli. | November | 26 | Helder donker | 1 | 3 | 29¼ |
| Uit de kleine Rhône om. trent Orgon. | November | 20 | Onklaer en door graeuw papier gefil. treerd | 1 | 3 | 29¼ |
| Te Cassis, Water uit 't Re gonbak van myn Labora torium. | January | 20 | Helder | 1 | 3 | 30 |

greinen 20

The past:

*The density of seawater measured by an early areometer (Marsigli, 1786 [1725])*

**A-7.2**

| | |
|---|---|
| **Variable:** | chlorinity |
| | **Cl** |
| **Unit:** | $10^{-3}$ |
| **Compartment:** | water |

**Introduction:**
Chlorinity is still used in estuarine research. The measurement is based on the constant ratio between the major constituents in seawater, including chloride. As chloride is usually not important in river water, this measurement is a better indication of mixing of fresh- and seawater, especially in the upper part of the estuary (at $S < 2 \times 10^{-3}$), where the interfering ionic composition of fresh water hampers the conductivity measurement.

**Sampling:**
Collection of water direct in a bottle, using a water sampler or a pump.

**Sample volume:**
1 litre

**Sample treatment:**
None

**Storage:**
Samples can be stored for several weeks preferably at reduced temperature in borosilicate glass bottles with screw caps (but not $< 0\ °C$, as during freezing precipitation will partly occur due to formation of micelles of different densities).

**Analytical methods:**
*Method #1.* Determination of chlorinity by titration of the chlorine with silver nitrate (Mohr-Knudsen titration) [1])

**Remarks:**
[1]) From chlorinity the salinity can be calculated according to the definition:
$S\ 10^{-3} = 1.80655\ Cl\ 10^{-3}$ (UNESCO, 1985).
Calibration of the chlorinity measurement should be performed regularly by using Standard Seawater (obtained from IAPSO Standard Sea Water Service).

**Method(s) references:**
Strickland & Parsons (1968; 1972); Grasshoff *et al.* (1983); Rodier (1984)

## TAFEL

*Thermometer, op verscheide dieptens, en op verscheide genomen, om deszelfs getemperdheid te onderzoeken.*

| Maenden | Dagen | Uuren | Toestand van den Thermometer op het water | Diepte des waters | Toestand van den Thermometer in de diepte der zee |
|---|---|---|---|---|---|
| | | | | Vademen | |
| December | 7 | 10 'sm | $9^{\frac{1}{2}}$ | 10 | $10^{\frac{3}{4}}$ |
| January | 18 | 10 'sm | $8\frac{1}{2}$ | 120 | $10\frac{1}{4}$ |
| January | 26 | 12 midd. | 9 | 20 | $10\frac{1}{4}$ |
| Maert | 28 | 10 'sm | 12 | 26 | $10\frac{1}{2}$ |
| Maert | 29 | 8 | 9 | 30 | $10\frac{1}{2}$ |
| April | 2 | 9 | $11\frac{1}{2}$ | 18 | $10\frac{1}{4}$ |
| Juny | 30 | 4 | 15 | 100 | 13 |
| Juny | 30 | 6 | 15 | 120 | 15 |

*The past:*
*Early thermometer for marine research, with an arbitrary scale (From: Marsigli, 1786 [1725])*

**A-7.3**

| Variable: | temperature |
| --- | --- |
| | T |
| Unit: | °C |
| Compartment: | water |

**Introduction:**
Temperature affects the rates of all chemical and biological processes. It also acts as a tracer for studies of the dispersion of thermal effluents.

Today either the semiconductor thermistor or the platinum resistance thermo-meter are most commonly used. Thermistors are cheaper but have an accuracy in the order of 0.1 °C, while a resistance thermometer has an accuracy of about 0.01 °C.

The mercury filled (deep-sea) thermometers are not often used in estuarine work, nor are the very accurate quartz crystal thermometers (Uncles *et al.*, 1983).

In-situ measurements are usually made by combined S and T (and depth) sensors in one probe. Vertical profiles are easily obtained and may indicate inhomogeneity or stratification in the water column.

**Sampling:**
*in situ* measurement at selected depths or along profiles whatever probe is used, regular calibration with high-precision calibrated thermometers in a stirred waterbath at different temperatures is necessary.

**Sample volume:**
non relevant

**Sample treatment:**
none

**Storage:**
non relevant

**Analytical methods:**
*Method #1a.* Thermistor
*Method #1b.* Resistance thermometer.
Usually the measurement is performed *in situ* using a Conductivity-Temperature-Depth (CTD) probe, or a more simple ST-meter, where one of the two sensor types is incorporated. Allow sufficient time to equilibrate at selected depths (5 sec), or lower the probe slowly (< 30 cm/s) when recording continuously.

*Method #2.* Thermometer. Traditional thermometers, filled with mercury or alcohol are known for their slow response. At least 30 sec should be allowed for equilibration. The oceanographic reversing thermometers seem to be a good option, but are rather tricky to use in shallow estuarine waters.

**Remarks:**

**Method(s) references:**
Aminot & Chaussepied (1983)

| | |
|---|---|
| **Variable:** | **turbidity** |
| | **Secchi depth, SD, Transmission, T$_r$** |
| **Unit:** | **Secchi disc: m, Transmission: %; Nephelometer: NTU** |
| **Compartment:** | **water** |

**Introduction:**

The transmission of light is affected by absorption and scattering. Because of the relatively high amount of suspended particulate matter in most estuaries, scattering is the most important process.

Determination of the 'Secchi depth' is a useful and simple way of providing an indication of the transparency of the (surface) water. More elaborate techniques involve the measurement of the transmittance of light over a given length (beam transmissometers) which can used either *in situ* or on board the sampling platform using a pumping system and flow through cell.

A more sophisticated type of instrument is the nephelometer, consisting of a collimator and a photocell set at an angle (usually 90°) to the light path and which measures directly the light scattered at a specific angle. *In situ* measurement is possible.

Data are reported in NTUs (nephelometric turbidity units) or FTUs (formazin turbidity units, formazin is a standard).

Optical back scattering (OBS) meters are in fact small nephelometers that integrate (infrared) radiation scattered at angles between 140° and 165°.

IR light does not penetrate very far in water thus the interference from sunlight is minimized. The major advantage of OBS is that the instruments have a good linearity, even in turbid waters. This is considered a great advantage in the highly variable estuarine environments. On the other hand, the advantage of transmissometers is, its better detection at low SPM concentrations, and the fact that the measured entity, attenuation, is an inherent optical property. For highly turbid waters a transmissometer is of little use, even when short light path lengths are used.

All types of measurements suffer from the problem that different types of suspensions scatter varying amounts of light, which means that estimation of the amount of suspended matter requires empirical calibration in the field (in space and time) by collection of samples [1].

Since turbidity is highly dependent on current speed, *in situ* measurements should be combined with current measurements. When a multi-sensor probe can be used, discrepancy between collection of data with several devices, and consequently measurement of different water masses, can be prevented (Fanger *et al.*, 1990).

**Sampling:**
Either *in situ* measurement (Secchi disc, Transmissometer, Nephelometer, Scatterometer) or on-board detection in a flow through cell under continuous pumping.
All methods allow the measurement of vertical profiles.

**Sample volume:**
not relevant

**Sample treatment:**
none

**Storage:**
not relevant

**Analytical method(s):**
All methods give specific results that cannot be compared between techniques; no ranking is implied here.
*Method #1.* Scatterometer. OBS is a rather small instrument that may be attached to a multi-sonde and thus lowered in the water column. Its large linear range in turbid waters makes them highly suitable for estuarine work. Calibration against standards and against actual samples is required (as for the other methods).
*Method #2.* Secchi depth. A white disc, 30 cm Ø, is lowered into the water.
The depth at which it is no longer visible is the 'Secchi depth' (SD). Disappearing and re-appearing should be checked.
*Method #3.* Beam transmission. The transmissometer is usually specific in design, with a fixed cell length (for estuarine waters between 1 and 10 cm).
The instrument gives a reading in percentage of full transmission (transmission in air or particle free water).
*Method #4.* Nephelometer. Either *in situ* or in the laboratory. As the scattering measured is measured under a 90° angle, nepholometry is more suitable for the detection of particles in low turbidity waters (transmittance > 90 %).

**Remarks:**
[1]   see the determination of the amount suspended particulate matter (SPM), procedure A-7.27.

**Method(s) references:**
Aminot & Chaussepied (1983); Morris (1983); Leatherland (1985); Instrument specific manuals

**Variable:**          **dissolved oxygen, DO**

                         **$O_2$**

**Unit:**               **mg/l [1)]**

**Compartment:**   **water**

**Introduction:**
The chemical determination of oxygen in seawater is traditionally done by the
Winkler titration, which is accurate and sensitive, and is used to check on other
methods of measurement.
New methods involve the use of oxygen sensing electrodes which are generally
of two types: galvanic or polarographic. These techniques measure the *in situ*
partial pressure of oxygen which is dependent on temperature, pressure and
salinity. Thus meters calibrated in % oxygen saturation will give accurate values
in both saline and freshwater, while those instruments with a concentration
readout will only be accurate in saline waters if a salinity correction has been
built-in in the instrument. Calibration thus becomes important; procedures are
given by HMSO (1980).
Other methods have been described (*e.g.* by mass spectrometry, gaschromato-
graphy, amperometry) but they have no practical use in estuarine research.

**Sampling:**
*In situ* analysis when applying electrodes, allowing ample time for equilibration
(> 30 sec). Vertical profiles are possible.
Collection of water using a water-sampler. Sub-samples for oxygen should be
collected as soon as possible after recovery of the sampler. Care must be taken
that no air is mixed with the water sample. This can be achieved by mounting a
clear plastic tube (no silicone tubing, as it is very permeable to oxygen) to the
outlet of the sampler, long enough to reach the bottom of the storage bottle. Take
care that no air is trapped in the tube and flush the bottle with sample water (at
least twice the volume), avoiding a too high turbulence; gently withdraw the tube
while the sample is still running. The bottle should be completely filled. Check
for air bubbles sticking to the wall, and remove them by gently tapping. Before
closure, reagents are added (see below).

**Sample volume:**
Typical sample volumes are in the order of 50-55 ml. Special, volume calibrated
glass (oxygen) bottles should be used, with special glass stoppers. Because of
calibration, the bottle and stopper should be numbered as they belong together.

**Sample treatment:**
For the Winkler titration the sample should be preserved immediately after collection. This is performed by adding manganese(II) chloride and alkaline iodide using a dispenser reaching almost the bottom of the flask and slowly withdrawn when the chemicals are added. The stopper is then inserted displacing about 5 ml of air-contaminated sample. Shake the bottles vigorously for one minute.

**Storage:**
During transport the stoppers may be secured with a rubber band. Ideally the analysis should be performed within 10-12 h of sampling. The bottles should be stored in the dark under even temperature conditions, preferably under water to minimize possible air exchange.

**Analytical method(s):**
Both Winkler titrations and the careful use of oxygen electrodes should yield the same order of accuracy.
*Method #1a.* Winkler titration. The dissolved oxygen is chemically bound by manganese(II) hydroxide in a strong alkaline medium. Mn(II) is oxidised to Mn(III) consuming the oxygen. Before titration the precipitated hydroxides are dissolved by acidification with sulphuric acid to pH < 2.5. Mn(III) oxidises iodide ions in an acidic medium to iodine, which is titrated with thiosulphate, using starch as an indicator (*i.e.* a colorimetric detection).
*Method #1b.* Oxygen electrode. For a detailed description of the working of the electrodes the manufacturers instructions should be followed. It should be noted that the slope will change over time and that calibration before each set of measurements is essential. Electrode membranes will have to be replaced from time to time, requiring full calibration afterwards. The electrodes should be stored under moist conditions. Lack of sufficient equilibration of the electrode in the sample is the most common source of error.

**Remarks:**
[1] both the units mg/l and % saturation are important. For the interpretation of estuarine processes the latter can be calculated if salinity and temperature are known.

**Method(s) references:**
Strickland & Parsons (1967; 1972); Aminot & Chaussepied (1983); Grasshoff *et al.* (1983); Parsons *et al.* (1984); Rodier (1984); Head (1985a)

**Variable:**        **pH**

**Unit:**             -

**Compartment:**   **water**

**Introduction:**
The pH, or the negative logarithm of the hydrogen ion activity (- log $a_{H^+}$) is today exclusively measured by an ion specific electrode (glass electrode), either as a glass electrode with a separate reference electrode, or a combined electrode device. The electrode is connected to a pH meter.
In-situ electrode systems are available, but often with a poor performance so far. Temperature and pressure affect the reading and care should be taken to correct for them. There is also a salt effect, but this is considered unimportant.
The care necessary for correct determination of pH is often underestimated, due to the false simpleness of the analysis (insert a stick in the sample, stir some-what and read three digits from the digital display).
The pH is one of the measurements that should preferably be made at the sampling location.

**Sampling:**
Because interaction with the $CO_2$ in the atmosphere will affect the pH in the sample, for very precise analysis the sample should be collected with the same care as for the oxygen measurements. When the samples cannot be analyzed directly, this procedure is essential. The subsampling from the water-sampler should be similar to the subsampling of oxygen. It involves mounting a clear plastic tube onto the outlet of the sampler, long enough to reach the bottom of the storage bottle. Take care that no air is trapped in the tube and flush the bottle with ample water (at least twice the volume), avoiding a too high turbulence; gently withdraw the tube while the sample is still running. The bottle is completely filled. Check for air bubbles sticking to the wall, and remove them by gently tapping.

**Sample volume:**
Glass bottles with glass stoppers, 50-100 ml should be used.

**Sample treatment:**
None

**Storage:**
Storage at reduced temperature, in the dark prior to measurement, which should be within 1 hour of sampling. Biological activity will seriously affect the pH.

**Analytical method(s):**
*Method #1.* Potentiometric determination of the hydrogen ion activity.
Measurement involves calibration with at least two buffer solutions, and at
controlled or at least recorded temperature. From the temperature during
measurement the pH *in situ* can be calculated (the field temperature should be
known).

**Remarks:**

**Method(s) references:**
Strickland & Parsons (1968; 1972); Aminot & Chaussepied (1983); Grasshoff *et
al.* (1983); Rodier (1984)

A-7.7

**Variable:** total alkalinity

$Alk_t$

**Unit:** mmol/l

**Compartment:** water

**Introduction:**
The alkalinity may be defined as the excess of anions of weak acids in seawater.
The alkalinity can be titrated and total alkalinity or titration alkalinity corres-
ponds to the amount of strong acid required to neutralize 1 kg of seawater.
Three different methods can be applied: the pH method, the back titration method
and the potentiometric titration method. Their results are not considered equally
accurate (precision respectively ±1, 0.2 and 0.8 %).

**Sampling:**
Samples can be obtained from any sampling device.

**Sample volume:**
Depending on the method and equipment used 100-500 ml in (brown) glass
bottles that have preferably been aged with hydrochloric acid for months. Bottles should be very clean to prevent/minimize bubble formation on the walls
around 'nuclei'.

**Sample treatment:**
If the samples cannot be analyzed soon after collection an exact amount of
hydrochloric acid should be added to the sample as preservative.

**Storage:**
Although exchange of $CO_2$ with the atmosphere does not hamper the analysis,
the loss of $CO_2$ by precipitation or biological activity does influence the analysis. Analysis should therefore be performed as soon as possible after collection.
See under sample treatment.

**Analytical method(s):**
*Method #1.* The pH method is a single point potentiometric titration of seawater.
The sample is acidified to about pH 3.5. The pH is measured with a high
precision pH meter and the alkalinity is calculated from the difference between
the amount of acid added and the excess acid present.
*Method #2.* The back titration method according to Gripenberg involves
acidification with hydrochloric acid to pH 3.5. Total $CO_2$ is driven off by boiling.

The solution is then back titrated with sodium hydroxide to pH 6 using an indicator and under purging with $CO_2$-free air.

*Method #3.* The potentiometric titration method is more complicated and involves iterative calculations and Gran plots. It requires the use of a computer for calculation. It involves the use of a pH electrode and a closed titration cell of accurately determined volume, under controlled or recorded temperature conditions.

**Remarks:**

**Method(s) references:**
Strickland & Parsons (1967; 1972); Grasshoff *et al.* (1983); Parsons *et al.* (1984); Rodier (1984)

**Variable:**  **nitrate**

  **$NO_3^-$**

**Unit:**  **μmol/l**

**Compartment:**  **water**

### Introduction:
Nitrate is the final oxidation product of nitrogen compounds in seawater and is considered the thermodynamically stable oxidation level under aerobic conditions. Nitrate is considered one of the most important nutrients controlling primary production. In estuaries large variations may occur, due to input and conversion in primary production. If the oxygen content becomes depleted as a result of microbial mineralization processes, nitrate may be used as an alternative electron acceptor. This denitrification happens in organic rich sediments and sometimes in estuarine waters (*e.g.* the Scheldt).
In the most frequently applied method, nitrate is reduced to nitrite, and the nitrite already present is corrected for. Manual and automated analytical methods (Autoanalyzer etc.) have been developed based on the same principle.

### Sampling:
Any sampling device is acceptable.

### Sample volume:
If all nutrients are sampled together, 0.5-1 l is more than sufficient. For nitrate the sample should be collected in glass bottles, or transferred to glass bottles as soon as possible (within 1 hour after sampling).

### Sample treatment:
In most (turbid) estuarine waters filtration should be applied (*e.g.* 0.45 μm, cellulose acetate). The first filtrate should be used to clean the filter, and is to be discarded.
No treatment is necessary, unless the samples have to be stored for a longer period of time. In that case ammonium chloride buffer or mercury chloride (0.01 % w/w) should be added as preservative.

### Storage:
Samples should be analyzed within 5 hours and, when no preservative is added, stored in the dark in a refrigerator. They may be stored deep frozen (- 20 °C) for several weeks. Chemically preserved samples should be stored dark and cool without refrigeration.

**Analytical method(s):**
*Method #1.* Nitrate reduction method. For this method the nitrate ions in the estuarine water, after buffering with ammonium chloride, pass over a reductor (copper coated cadmium granules), and are quantitatively reduced to nitrite. Detection is similar to the detection of nitrite through an azo dye, which is spectrophotometrically detected (see A-7.9). The original nitrite should be corrected for.

**Remarks:**

**Method(s) references:**
Strickland & Parsons (1967; 1972); Aminot & Chaussepied (1983); Grasshoff *et al.* (1983); Parsons *et al.* (1984); Rodier (1984); Head (1985a); Kirkwood *et al.* (1991)

**A-7.9**

| | |
|---|---|
| **Variable:** | **nitrite** |
| | $NO_2^-$ |
| **Unit:** | µmol/l |
| **Compartment:** | water |

**Introduction:**
Nitrite forms the intermediate between reduction of nitrate or the oxidation of ammonia in sea- and estuarine water. Usually the concentrations are rather low, except in transition zones between oxic and anoxic conditions. Especially in polluted estuaries it can become an important aspect of the nitrogen cycle. Only one spectrophotometric detection method is commonly used.
Manual and automated analytical methods (Autoanalyzer etc.) have been developed based on the same principle.

**Sampling:**
Any sampling device is acceptable.

**Sample volume:**
If all nutrients are sampled together, 0.5-1 l is more than sufficient.

**Sample treatment:**
In turbid waters filtration over a prewashed polycarbonate 0.4 µm Nuclepore or 0.45 Ø membrane filter is necessary. Because of biological (bacterial) activity, analysis should be performed within 30 min of sampling, or at least the reagents should have been added. Mercury chloride (0.01 % w/w) may be added as preservative.

**Storage:**
Samples (with the reagents added) should be analyzed within 5 hours. Direct sunlight should be avoided. Samples may be stored deep frozen (- 20 °C), though prolonged storage is not recommended.
Chemically preserved samples should be stored dark and cool without refrigeration.

**Analytical method(s):**
*Method #1.* The nitrite in sea- and estuarine water is allowed to react with sulphanilamide in an acid solution. The resulting diazo compound reacts with N-(1-naphthyl)-ethylenediamine and forms a highly coloured dye. The extinction is measured spectrophotometrically at 540 nm.

**Remarks:**

**Method(s) references:**
Strickland & Parsons (1967; 1972); Aminot & Chaussepied (1983); Grasshoff *et al*. (1983); Parsons *et al*. (1984); Rodier (1984); Head (1985a)

**Variable:**         **ammonia**

                      **$NH_4$**

**Unit:**             µmol/l

**Compartment:**      water

## Introduction:

In all methods the sum of $NH_3$ and $NH_4^+$ is determined, which is not unrealistic because of the acid-base pair that exists in nature ($NH_4^+$-$NH_3$).
Two major methods are used which, however, measure different amounts of
nitrogen. The oxidation method determines not only the ammonia but also a
considerable fraction of the amino acids (which may not be undesirable in
production studies, as the amino acids will be used as an N-source as well). The
alternative more preferable method determines ammonia only. More methods
can be found in the literature, however. Manual and automated analytical
methods (Autoanalyzer etc.) have been developed based on the same principle.

## Sampling:

Any sampling device is acceptable.

## Sample volume:

If all nutrients are sampled together, 0.5-1 l is more than sufficient.

## Sample treatment:

No treatment required. Filtration is not recommended because of the serious
chance of contamination by the filters.

## Storage:

Samples for the determination of ammonia should be analyzed as soon as
possible after collection. Storage in a refrigerator is possible up to 3 hours.
Samples can be stored for up to 2 weeks after addition of phenol as a preservative, or freezing at - 20 °C.

## Analytical method(s):

*Method #1.* Ammonia. The sample is treated in an alkaline citrate medium
with sodium hypochlorite and phenol in the presence of sodium nitroprusside
which acts as a catalyst. The blue indophenol colour formed with ammonia is
measured spectrophotometrically at 640 nm.
*Method #2.* Oxidation method. The ammonia (and some other amino compounds) in sea and estuarine water is oxidized to nitrite by alkaline hypochlorite

at room temperature and the excess oxidant destroyed by the addition of arsenite. The nitrite is determined spectrophotometrically by the procedure given under A-7.9.

**Remarks:**
Avoid smoking and other air-borne contamination during sample handling.

**Method(s) references:**
Strickland & Parsons (1967; 1972); Aminot & Chaussepied (1983); Grasshoff *et al*. (1983); Parsons *et al*. (1984); Rodier (1984)

**A-7.11**

| | |
|---|---|
| **Variable:** | **phosphate** |
| | **$PO_4^{3-}$** |
| **Unit:** | **µmol/l** |
| **Compartment:** | **water** |

**Introduction:**
Phosphorus is one of the most important nutrients. Inorganic phosphate is present in seawater in the form of ions of (ortho) phosphoric acid, $PO_4^{3-}$ and $HPO_4^{2-}$. In addition to these (and other possible) inorganic forms, organic phosphorus compounds are present. Speciation into various phosphorus species is possible, as is the determination of total phosphorus, but this is beyond the scope of this manual. Generally only one method is used for the detection of inorganic phosphate.
Manual and automated analytical methods (Autoanalyzer etc.) have been developed based on the same principle.

**Sampling:**
Any sampling device is acceptable.

**Sample volume:**
If all nutrients are sampled together, 0.5-1 l is more than sufficient. For phosphate the sample can be collected in polythene or glass bottles, or transferred to glass bottles.

**Sample treatment:**
Filtration can cause serious errors, because of contamination of the sample. The use of 0.4 µm Nuclepore filters has been recommended, after washing with part of the sample water.

**Storage:**
Preferably samples are analyzed directly after sampling, within 2 hours, but if necessary may be stored for a short period in a cool dark place. Several methods have been suggested for long term storage: deep frozen (at - 20 °C), addition of sulphuric acid or of chloroform. Mercury chloride (0.01 % w/w) has been suggested as preservative. There is conflicting evidence on the effects of these storage methods, however.

**Analytical method(s):**
*Method #1.* The sample is allowed to react with a composite reagent containing molybdic acid, ascorbic acid and trivalent antimony. The resulting complex acid is reduced *in situ*, and measured spectrophotometrically at 885 nm.

**Remarks:**

**Method(s) references:**
Strickland & Parsons (1967; 1972); Aminot & Chaussepied (1983); Grasshoff *et al*. (1983); Parsons *et al*. (1984); Rodier (1984); Head (1985a)

**A-7.12**

---

| **Variable:** | **silicate** |
| | **$H_4SiO_4$** |
| **Unit:** | **μmol/l** |
| **Compartment:** | **water** |

**Introduction:**
Silicon is an essential element especially for diatoms. It is present in seawater in the form of orthosilicic acid $Si(OH)_4$. Usually a method involving a silicomolyb-dic complex is applied which, however, only measures the so called "reactive" silicate. Several routes can be followed. Manual and automated analytical methods (Autoanalyzer etc.) have been developed based on the same principle.

**Sampling:**
Any sampling device can be used.

**Sample volume:**
If all nutrients are sampled together, 0.5-1 l is more than sufficient. For silicates the sample should never be collected in glass bottles, but in polythene or PVC bottles.

**Sample treatment:**
Filtration over a 0.45 μm membrane filter is recommended in estuarine waters as the (re)suspended silicate rich particulate matter will overestimate the dissol-ved silicate content. No further treatment is required, unless prolongued storage is espected. Preservation by addition of sulphuric acid to pH 2.5, or addition of mercury chloride (0.01 % w/w) as preservative is then recommended.

**Storage:**
Storage of a sample in the refrigerator in the dark does not change the silicate content significantly. Polymerisation of orthosilicate has been reported for freshwater and may be important in the upper estuary. Storage at - 20 °C can be used provided that the samples are left for at least 3 hours after thawing.

**Analytical method(s):**
*Method #1a.* The estuarine water sample is allowed to react with molybdate under conditions which result in the formation of silicomolybdate, phosphomo-lybdate and arsenomolybdate. A metol and oxalic acid containing solution is ad-ded which reduces the silicomolybdate complex (or to give a blue colour (detec-tion at 810 nm) and at the same time decomposes the other two molybdates.

*Method #1b.* The acidified sample is treated with a molybdate solution to form yellow coloured silicomolybdic acid, which can be detected at 390 nm (ß-isomeric form) after addition of oxalic acid and ascorbic acid, the latter as reductant.

**Remarks:**

**Method(s) references:**
Strickland & Parsons (1967; 1972); Aminot & Chaussepied (1983); Grasshoff *et al*. (1983); Parsons *et al*. (1984); Rodier (1984); Head (1985a); Eberlein & Kattner (1987)

**Variable:**      **hydrogen sulphide (and other sulphides HS⁻, S²⁻)**

$H_2S$

**Unit:**           µmol/l

**Compartment:**   water

**Introduction:**
When the oxygen is depleted by bacterial mineralisation processes, sulphate is used as an electron donor and consequently sulphide is produced. In these anoxic waters, that may also occur in estuaries when a high load of organic matter (sewage) is introduced in the system, the sulphide is dissolved in a form of $H_2S \leftrightarrow HS^- \leftrightarrow S^{2-}$ depending on the pH. Sulphides ($S^{2-}$) are particularly reactive towards (heavy) metals.
Two colorimetric methods, the methylene blue method and Lauth's violet method, are used. A titrimetric method has also been described, but is less accurate. The latter method is not exclusive for the detection of $H_2S$ and will also include effects of other reductants.

**Sampling:**
The oxygen in the air will readily oxidize the hydrogen sulphide. Therefore similarly careful sampling as in the case of dissolved oxygen has to be applied. Plastic or glass water samplers should be used as metal parts influence the sulphide content.
Sub-samples for hydrogen sulphide should be collected as soon as possible after recovery of the sampler. Care must be taken that no air is mixed with the water sample. This can be performed by mounting a clear plastic tube onto the outlet of the sampler, long enough to reach the bottom of the storage bottle. Take care that no air is trapped in the tube and flush the bottle with sample water (at least twice the volume), avoiding a too high turbulence; gently withdraw the tube while the sample is still running. The bottle is completely filled. Check for air bubbles sticking to the wall, and remove them by gently tapping.

**Sample volume:**
Typical sample volumes are in the order of 50-55 ml. Special glass (oxygen) bottles should be used with special glass stoppers.

**Sample treatment:**
No treatment is required (see under storage)

**Storage:**
Samples may be preserved by addition of zinc acetate or -chloride, which results in the precipitation of ZnS. When kept in the dark the samples can thus be stored for a long time.

**Analytical method(s):**
*Method #1a.* Addition of dimethyl-p-phenylene diamine and ferric chloride results in the formation of methylene blue which is detected spectrophoto-metrically at 670 nm.
*Method #1b.* To the acidified sample p-phenylene diamine and ferric chloride are added, resulting in the formation of Lauth's violet, which is detected spectrophotometrically at 600 nm.
*Method #2.* The hydrogen sulphide in the sample is precipitated with manganous sulphate reagent and alkaline potassium iodide (see Winkler titration, in A-7.5). Manganous sulphide is formed and precipitated with the hydroxide. Addition of potassium iodate and sulphuric acid, and titration with thiosulphate using starch as an indicator.
*Method #3.* Ion specific electrode (ISE). This electrode measures the form $S^{2-}$ only. To calculate the total amount of sulphides ($H_2S \leftrightarrow HS^- \leftrightarrow S^{2-}$) the pH and $pK_1$, $pK_2$ should be taken into consideration.

**Remarks:**

**Method(s) references:**
Strickland & Parsons (1967; 1972); Grasshoff *et al.* (1983); Rodier (1984)

**Variable:**        **sulphate**

                **$SO_4^{2-}$**

**Unit:**        **µmol/l**

**Compartment:**    **water**

**Introduction:**
Sulphate is abbundant in the marine environment, and in seawater a constant sulphate/chlorinity ratio can be found. Per kg standard seawater of salinity 35 x $10^{-3}$ 2.773 g (0.02889 M) sulphate is present. In estuaries anoxic water conditions are not uncommon, due to the mineralisation of organic matter. Under these conditions sulphate is reduced to sulphide(s) by sulphate reducing bacteria. Thus in estuaries the constant ratio may change.
Analysis can be performed in various ways, the most important include gravimetric and turbidimetric methods. Also the methylene blue method that is used for the determination of $H_2S$ (see A-7.13) has been applied, after reduction of sulphate. This method has been adapted for use in an Autoanalyzer, involving methylthymol blue. Radiochemical (use of [131]Ba or [133]Ba), atomic absorption (AAS, detection of Ba) and ion chromatographic methods exist, but they will not be treated here.

**Sampling:**
Any sampling device is acceptable.

**Sample volume:**
If all nutrients are sampled together, 0.5-1 l is more than sufficient. The sample can be collected in polythene or glass bottles.

**Sample treatment:**
The filtration of the watersample over a 0.2 µm filter is recommended to remove bacteria that may interfere (see storage).

**Storage:**
In the presence of organic matter bacteria may reduce sulphate to sulphide. To prevent this the samples should be stored deep froozen at - 20 °C.

**Analytical method(s):**
*Method #1.* Tubidimetric determination. The sulphate in the sample is quantitatively precipitated as small, uniformly sized crystals of $BaSO_4$ in a suspension medium containing HCl and gelatin of which the turbidity is determined on a colorimeter fitted with a blue filter.

*Method #2.* The gravimetric method also involves the quantitative precipitation of sulphate in the sample as $BaSO_4$ that is formed after addition of an excess of barium chloride. The precipitate is washed, dried and weighed.

*Method #3.* Barium sulphate is formed by the reaction of the $SO_4^{2-}$ with barium chloride at low pH. At high pH excess barium reacts with methylthymol blue to produce a blue chelate. The uncomplexed methylthymol blue is gray. The detection of this uncomplexed indicator is by colorymetry using a 460 nm filter.

*Method #4.* After reduction of the sulphate to sulphide, the sulphide is colorimetrically detected by the methylene blue method (see A-7.13).

**Remarks:**

One should be aware of the processes that may lead to the transformation of either sulphate into sulphide by bacteria, or of sulphide into sulphate due to (rapid oxidation in air), especially in organic matter rich estuaries.

**Method(s) references:**

Johnson & Nishita (1952); Berglund & Sörbo (1960); Morris & Riley (1966); Tabatabai (1974); APHA *et al.* (1985)

**A-7.15**

| | |
|---|---|
| **Variable:** | **dissolved organic carbon** |
| | **DOC** |
| **Unit:** | **mg/l** |
| **Compartment:** | **water** |

**Introduction:**
Carbon is present in sea- and estuarine waters in organic and inorganic forms.
The latter occur in concentrations that are orders of magnitude higher than the
DOC. Particulate and dissolved organic carbon play a very important role in the
estuarine system, as they contribute to a large extent to the oxic/anoxic
conditions, the biological activity etc. (Duursma & Dawson, 1981).
Several methods have been described and are used, all based on the quantitative
destruction of the DOC and subsequent measurement.
The methods involve either chemical wet oxidation, high temperature combustion (with or without a catalyst), and photo-oxidation.
Unfortunately there is strong evidence that not all methods break down the DOC
quantitatively, and some refractory material remains undetected (Sugimura &
Suzuki, 1988). How far this discrepancy is important for estuarine work is as
not clear yet. As long as there is no general agreement upon the use of one
technique, it seems unrealistic to prescribe one method as the preferential
choice. Nevertheless the better the breakdown of organic matter, the better the
technique, despite the possibility that some refractory material is not available
for organisms. The sequence of the methods listed here follow more or less a
decrease in their ability to destroy dissolved organic matter.
In some methods there may be serious interference from the chlorine that
develops during destruction. Moreover, it is necessary to avoid the interference
of inorganic carbon. The methods used for this step still cause many problems.
Manual and automated analytical methods (Autoanalyzer etc.) have been
developed based on some of these principles. Specialized instruments have been
produced.

**Sampling:**
Because of the very low DOC concentrations in seawater great care must be
taken to avoid contamination. Oil and grease from sampling gear is a major
source of error. Plastic water samplers or glass samplers can be used. For surface water sampling, care must be taken not to collect the organic rich surface
micro layer.

**Sample volume:**

The collection of 1 litre water in (dichromate/sulphuric acid) cleaned glass bottles with glass stoppers allows later filtration and collection of particulate organic matter. A 100 ml sample is sufficient for DOC analysis only.

**Sample treatment:**

In estuarine waters it is essential to filter the samples prior to analysis. Pre-ignited (450 °C overnight) glass fibre filters should be used (GF/C or GF/F).

**Storage:**

The method of storage will partly depend on the method of analysis. Biological activity must be prevented. This has traditionally been performed by addition of mercury chloride (0.01 % w/w) as a preservative; this is a serious toxic agent, however, and is banned in several institutes. It may also interfere with the analysis, especially where catalysts are involved. A minimum requirement is cold storage in the dark.

Addition of phosphoric acid to pH < 4.5 has been proposed, but suffers from potential contamination in waters with a low DOC concentration.

In methods which use ampoules, the samples, once sealed can be stored for a long time, especially after they have been autoclaved.

**Analytical method(s):**

*Method #1.* High temp. catalytic oxidation. The oxidation is carried out on a platinum catalyst at 680 °C under an oxygen atmosphere after the sample has been freed of inorganic carbon. The concentration of the $CO_2$ is measured with a non-dispersive IR gas analyser.

*Method #2.* High temp. catalytic oxidation. The nitric acid acidified sample is freed from inorganic and $CO_2$ volatile organic compounds. The liquid phase is introduced in the copper filled oxidation reactor maintained at 880 °C. Collected $CO_2$ is desorbed, mixed with hydrogen and passed through a methane reactor containing a nickel catalyst. The methane formed is detected using a flame ionisation detector (FID).

*Method #3.* High temp. catalytic oxidation. The nitric acid acidified sample is freed from inorganic and $CO_2$ volatile organic compounds. The liquid phase is introduced a nickel oxide filled oxidation reactor maintained at 1000-1200°C. Collected $CO_2$ is desorbed, mixed with hydrogen and passed through a methane reactor containing a nickel catalyst. The methane formed is detected using a flame ionisation detector (FID).

*Method #4.* Wet chemical oxidation. Phosphoric acid and $K_2S_2O_8$ (persulphate) are added to the sea-water, the inorganic $CO_2$ is purged out with a $CO_2$-free gas and the sample is sealed in ampoules. The DOC is oxidized by heating the ampoules in an autoclave, the resulting carbon dioxide measured by infrared analyzer.

*Method #5.* UV oxidation. The acidified sample is purged with nitrogen to remove carbonates. After a persulphate/UV destruction of the DOC in a quartz coil the $CO_2$ is dialysed into a (bi)carbonate-phenolphthalein colour reagent and photometrically detected at 550 nm.

*Method #5.* Dry combustion. The method involves the evaporation of the water and burning the residue in a combustion tube. The method suffers from high blanks.

**Remarks:**

**Method(s) references:**
Gershey *et al.* (1979); MacKinnon (1981); Grasshoff *et al.* (1983); Rodier (1984); LeB-Williams (1985); Sugimura & Suzuki (1988); Toggweiler (1988, 1989); Suzuki *et al.* (1992); Specific instrument manuals

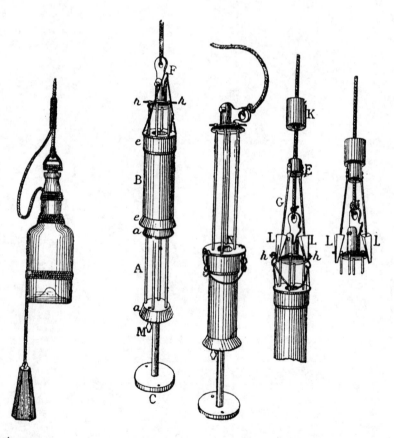

*The past:*
*Water sampling bottles of the 'Kiel Commission' (l, 1873) and of Meyer (r, 1889)*
*(From: Thoulet, 1890)*

**A-7.16**

| Parameter: | **dissolved organic nitrogen** |
| --- | --- |
| | **DON** |
| Unit : | **mg N/l** |
| Compartment: | **water** |

**Introduction:**
Several methods for the analysis of DON have been described, including high temperature combustion with a catalyst, some wet oxidation methods and photo destruction by UV light. There are indications that some of the methods, especially the photo-oxidation method, suffer from incomplete oxidation of organic nitrogen compounds. Nevertheless it is still under discussion whether analytical methods should cover all nitrogen compounds without regard to their origin and fate, or should detect only the biologically accessible compounds. The analytical problems appear to be similar to those observed for the DOC analysis.

**Sampling:**
Because of the very low DON concentrations in seawater great care must be taken to avoid contamination. Plastic water samplers or glass samplers can be used. For surface water sampling, care must be taken not to collect the organic rich surface micro layer.

**Sample volume:**
The collection of 1 litre water in (dichromate/sulphuric acid) cleaned glass bottles with glass stoppers allows later filtration and collection of particulate organic matter. A 100 ml sample is sufficient for DON analysis.

**Sample treatment:**
In estuarine waters it is essential to filter the samples prior to analysis. Pre-ignited (450 °C overnight) glass fibre filters should be used (GF/C or GF/F).

**Storage:**
The method of storage strongly depends on the method of analysis. Generally, biological activity must be prevented. This could be achieved either by addition of mercury chloride (0.01 % w/w) as a preservative or by cold storage in the dark. In the case of wet chemical analysis, the samples should be stored in (wet) oxidized form.

**Analytical method(s):**
*Method #1.* Persulphate method. DON will be analyzed after oxidation with potassium peroxidisulphate (persulphate) with boric acid and sodium hydroxide

at 125 °C under pressure in a destruction bomb. A pressure cooker may be used instead. After cooling the organic nitrogen has been converted into nitrate which is analyzed according to the method for nitrate as given under A-7.8.

*Method #2.* Micro-Kjeldahl method. Total Kjeldahl nitrogen in the sample is determined by digestion of the organic nitrogen in an aqueous sulphuric acid solution. The ammonia produced is determined in the neutralized digest by a colorimetric method.

*Method #3.* High temperature catalytic oxidation method. The nitrogenous sample is oxidized on a platinum catalyser at 680 °C under an oxygen atmosphere. The $NO_2$ thus generated is absorbed in a chromogenic reagent, followed by spectrometric detection.

*Method #4.* UV oxidation. The sample, contained in quartz glass tubes (or coil) is treated with UV light from a mercury arc tube to decompose the organic matter. The nitrogen compounds are oxidized to nitrate which is measured according to A-7.8.

**Remarks:**

Mercury chloride is a toxic compound. Please refer to the safe handling procedures for chemicals.

**Remarks:**

**Method(s) references:**

Parsons *et al.* (1984); Suzuki & Sugimura (1985)

**Variable:**        **dissolved humic compounds (yellow acids)**

**Unit:**        **mFl**

**Compartment:**    **water**

**Introduction:**
Dissolved humic compounds are a largely uncharacterized class of organic
compounds, which are assumed to be generally of terrigenous origin. They can
also be produced in the aquatic environment, however. They show a conservative
behaviour in coastal marine waters as was demonstrated by studies concerning
estuarine mixing processes (Zimmerman & Rommets, 1974; Laane, 1981).
Concentrations of dissolved humic compounds or yellow substance ('Gelbstoff',
Kalle, 1937) are usually determined by the detection of the natural fuorescence.
Kalle (1963) defined 'milli-fluorescence', based on the fluorescence of quinine
bisulphate and expressed as mFl/l. By definition, 1 mg quinine bisulphate in
0.005 M $H_2SO_4$ has a fluorescence of 700 mFl.
The fluorimetric determination of the concentration of dissolved humic
compounds offers the possibility of continuous measurement, thus providing a
useful tool directly on board the ship for monitoring and rapid distinguishing
water masses of different origin. *In situ* measurements using a probe are in
principle possible.

**Sampling:**
As for water (S-6.1). Avoid contamination of the sample, *e.g.* by sampling the
surface microlayer.

**Sample volume:**
0.5 to 1 litre.

**Sample treatment:**
For specialised measurements, the sample is filtered over a pre-combusted
(550 °C, overnight) GF/C or GF/F glass fibre filter. However, the method
operates well without filtration in waters with low turbidity.

**Sample storage:**
Direct measurement is preferred. Water samples may be stored deep frozen at
- 18 °C.

**Analytical method(s):**
*Method #1. Filter fluorescence measurements* on a filtered sample using a
fluorimeter. The excitation wavelength is 358 nm (at maximum transmission,

primary filter *e.g.* type 7-60), the emission wavelength is 457 nm (dito, secondary filterset of a Wratten 48 and 2A).

Fluorimetric readings may be corrected for self-absorption of the excited and emitted light according to Duursma & Rommets (1961).

*Method #2. Fluorescence scan measurements* on a filtered sample using a scanning fluorimeter. The excitation wavelength is 320-365 nm, the emission wavelength can be varied from 350 to 550 nm.

**Remarks:**

In method #2 Raman scattering may interfere and overlap with the peaks in the fluorescence spectrum, thus hampering the accuracy of the determination. However, for routine measurements quantum correction of the emission spectra does not appear to be feasible, because of the sophisticated techniques involved.

**Method(s) references:**

Kalle (1937, 1963); Zimmerman & Rommets (1974); Laane (1981); Liebezeit (1988); Laane & Kramer (1990)

**A-7.18**

| | |
|---|---|
| **Variable:** | **total carbohydrates** |
| **Unit:** | **µg C/l or µg C/g** |
| **Compartment:** | **water, seston, sediment** |

**Introduction:**
Carbohydrates, or better total carbohydrates, are part of the menu of estuarine bacteria and plankton, and constitute of an important, easily digestable carbon source. There are indications that carbohydrates are converted passing the estuary and that they are essential for the formation of particles and the flocculation process.
Analysis can be separated into the dissolved and total fraction, of which the analysis, apart from filtration is the same.
Automated analysis has been developed.

**Sampling:**
As for water (S-6.1). Care should be taken not to contaminate the sample. Cleaning procedures as for DOC (A-7.15). Avoid sampling of the surface micro-layer.

**Sample volume:**
One litre is more than sufficient.

**Sample treatment:**
To separate the dissolved from the particulate carbohydrates, the sample is filtered over a pretreated GF/C or GF/F glass fiber filters in precleaned glassware.

**Storage:**
Storage in the refrigerator at 4 °C after addition of mercury chloride (0.01% w/w) for conservation is recommended.

**Analytical method(s):**
*Method #1.* After filtration the filter is treated with 10 ml 1 M sulphuric acid at 100 °C. Either the dissolved fraction or the (treated) particulate phase is mixed with a reagent containing tryptophan, boric acid and sulphuric acid, and heated for 15 min at 100 °C. The resulting violet coloured complex is measured photometrically at 520 nm. Nitrate gives a brownish complex and may at high concentrations disturb the carbohydrate analysis.

**Remarks:**

**Method(s) references:**
Eberlein & Hammer (1980); Dawson & Liebezeit (1981); Eberlein & Schütt (1986)

**A-7.19**

---

**Variable:**      **individual carbohydrates**

**Unit:**          **µg C/l or µg C/g**

**Compartment:**   **water, seston, sediment**

**Introduction:**
For the analysis of mono- and oligosaccharides the procedures for detection in
the dissolved phase or in the seston are the same, except for the filtration step.
Methods include the identification of individual (mono)saccharides (Dawson &
Liebezeit, 1981).

**Sampling:**
As for water (S-6.1). Care should be taken not to contaminate the sample.
Cleaning procedures as for DOC (A-7.15). Avoid sampling of the surface
micro-layer. Collection in cleaned glass containers.

**Sample volume:**
200 ml.

**Sample treatment:**
To separate the dissolved fraction from the seston, the sample is filtered over a
GF/C or GF/F glass fiber filter.

**Storage:**
Storage at 4 °C after preservation by addition of mercury chloride (0.01 % w/w)

**Analytical method(s):**
*Method #1.* For the preparation of the combined monosaccharides or the
particulate saccharides, the samples that are reduced in size by evaporation,
following desalting of the filter respectively are hydrolysed using 2 M HCl in a
sealed glass tube for 3.5 h at 100 °C.
Depending on the salt content of the filtered sample, the salt is removed by
electrodialysis for 2-8 h (for particulate matter 15 min is appropriate) using an
ion exchange membrane and 60-200 V DC.
The dialysed sample is reduced in volume by a rotation evaporator.
The thus purified sample is eluted by HPLC using a highly alkaline anion
exchange resin in a basic borate buffer. The individual mono-/oligomeres are
detected as ethylene diamine complexes using a fluorimeter (excitation 320 nm,
emission 460 nm).
*Method #2.* The sample is derivatized with methanolic 1-phenyl-3-methyl-5-
pyrazolon and aqueous sodium hydroxide at 70 °C. After neutralization and

washing with chloroform the aqueous solution is eluted with acetonitrile/
phosphate buffer by reversed phase chromatography using a $C_{18}$ column. The
mono/oligomere derivatives are detected by UV spectrometry (at 245 nm) or by
electrochemical method (at 600 mV potential).

*Method #3.* Free monosaccharides are reduced to sugar alcohols with $KBH_4$,
periodate oxidation and subsequent spectrometric determination of the liberated
formaldehyde with 3-methyl-2-benzothiazolinone hydrazone hydrochloride.
This gives the monosaccharide content. The poly-saccharide content is
determimined after a hydrolysis step using 0.1 M HCl.

**Remarks:**
Method #3 is quite unspecific; alditols, amino sugars and uronic acids are
detected simultaneously.

**Method(s) references:**
Johnson & Sieburth (1977); Mopper (1977, 1978, 1980); Dawson & Liebezeit
(1981); Honda *et al.* (1989)

**A-7.20**

---

**Variable:**        **total amino acids**

**Unit:**            **µg N/l or µg N/g**

**Compartment:**     **water, seston, sediment**

**Introduction:**
Dissolved amino acids pass, by definition, a 0.45 µm filter; all amino acids
together that pass through are the total amino acids.
The method of fluorimetric analysis reported here determines total amino acids in
the water phase, rather than single amino acids. Individual amino acids can
be determined after chromatographic separation.

**Sampling:**
As for water (S-6.1). Care should be taken not to contaminate the sample.
Cleaning procedures as for DOC (A-7.15). Avoid sampling of the surface
micro-layer.

**Sample volume:**
One litre.

**Sample treatment:**
The sample is filtered over a precleaned GF/C or GF/F glass fiber filter.

**Storage:**
Storage in the refrigerator at 4 °C after addition of mercury chloride (0.01% w/w)
for conservation.

**Analytical method(s):**
*Method #1.* Fluorimetry. After addition of a reagent consisting of o-phtalal-
dehyde/mercaptophenol in an alkaline borate buffer, the filtered sample is
incubated at room temperature. The resulting fluorescent complex is measured fluorimetricaly (excitation 356 nm, emission 450 nm).

**Remarks:**
Proteins, urea and/or ammonia at higher concentrations disturb the analysis,
which might be eliminated by gel filtration prior to further analysis.

**Method(s) references:**
Dawson & Liebezeit (1981); Hammer & Luck (1987)

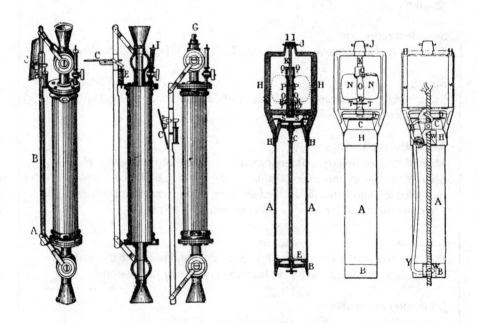

*The past:*
*Water sampling bottles of Buchanan (l, 1893) and of Sigsbee (r, 1880) (From: Thoulet, 1890)*

**Variable:**        **individual amino acids and amino sugars**

**Unit:**        **µg/l or µg/g**

**Compartment:**   **water, seston, sediment**

**Introduction:**
As a major constituent of plant, animal and microbial biomass, amino acids and amino sugars are one of the main sources of organic matter in the aquatic environment. The can be used as biogeochemical indicators, as their monomeric composition as well as their monomeric ratios provide information about the sources of the organic matter. Changes in the monomeric amino acids and -sugar composition additionally allow one to describe the degradation and transformation processes of the dissolved and particulate material.

**Sampling:**
As for water (S-6.1). Avoid contamination of the sample, *e.g.* by sampling the surface microlayer.

**Sample volume:**
0.5 to 1 litre.

**Sample treatment:**
The sample is filtered over a pre-combusted (550 °C, overnight) GF/C or GF/F glass fibre filter. The filter should be rinsed with distilled watr to remove salt.

**Sample storage:**
Water samples are stored at 4 °C in the refrigerator after addition of mercury chloride (0.01 % w/w) for preservation. Seston and sediment should be freeze-dried. Alternatively, they should be dried at 40 °C or stored deep frozen at - 20 °C until analysis.

**Analytical method(s):**
*Method #1*. Analysis of individual amino acids and sugars is performed on hydrolysed samples of known volume (water) or weight (sediment, seston). The hydrolysis is carried out under an argon atmosphere in a pre-combusted glass ampoule with 6 M HCl at 110 °C for 22 h. After hydrolysis, a known volume of the particle free hydrolysate is evaporated to dryness and washed twice with double distilled water to remove residual HCl. The dried residue is dissolved in a citrate buffer (pH 2.1), from which aliquots can be analysed by liquid chromatography using an anion exchange resin with sodium citrate as eluents. Individual amino acids are quantitatively measured fluorimetrically (external standard) after post column addition of a reagent consisting of OPA/mercapto-

ethanol in an alkaline borate buffer. For measurement of free dissolved amino acids and -sugars the filtered hydrolysed sample can be analysed directly.
*Method #2.* The hydrolisate (see Method #1) is analysed fluorimetrically using a pre-column derivatization of the hydrolysate in an alkaline buffer (0.1 M, pH 9.5) on a HPLC device with a 'sperisorb' analytical column at room temperature. A phosphate buffer (0.02 M, pH 5.6) and methanol are used as eluents.

**Remarks:**
Mercury chloride is a toxic compound. Please refer to the safe handling procedures for chemicals.

**Method(s) references:**
Garassi *et al*. (1979); Lee & Cronin (1982, 1984); Ittekot *et al.* (1984); Haake *et al.* (1992)

**A-7.22**

---

**Variable:** proteins

**Unit:** µg N/l or µg N/g

**Compartment:** water, seston, sediment

**Introduction:**
Particulate and dissolved proteins follow the same analytical procedure, with the exception of the filtration step.

**Sampling:**
As for water (S-6.1). Glass or plastic samplers can be used. Care should be taken not to contaminate the sample. Cleaning procedures as for DOC (A-7.15). Avoid sampling of the surface micro-layer. Collection in cleaned glass containers.

**Sample volume:**
One litre.

**Sample treatment:**
No treatment is performed for the dissolved proteins. To separate the dissolved phase from the suspended particulate matter, a precleaned GF/C or GF/F glass fiber filter is used.

**Storage:**
Storage in the refrigerator at 4 °C after addition of mercury chloride (0.01% w/w) for conservation.

**Analytical method(s):**
*Method #1.* The filter is hydrolysed using 2 M NaOH, which is followed by centrifugation
The 'extract' or the filtered water are measured photometrically as brilliant-blue complex in a phosphoacıdic-methanolic solution at 625 nm, using bovine protein as a standard.
*Method #2.* After gel-filtration, separating the low molecular weight substances, the same fluorimetric method is applied that is described for the analysis of amino acids (A-7.20).

**Remarks:**
Higher concentrations of amino acids may disturb the analysis; a gel filtration separation step must then be included in the sample treatment. Care must be taken to avoid adsorption of proteins to surfaces.

**Method(s) references:**
Dawson & Liebezeit (1981); Hammer & Nagel (1986)

**A-7.23**

**Variable:**          **lipids**

**Unit:**               **µg C/l or µg C/g**

**Compartment:**   **water, seston, sediment**

**Introduction:**
Dissolved and particulate lipids follow the same analytical procedure, except
for the filtration step.

**Sampling:**
As for water (S-6.1). Care should be taken not to contaminate the sample.
Cleaning procedures as for DOC (A-7.15). Avoid sampling of the surface
micro-layer. Collection in cleaned glass containers.

**Sample volume:**
One litre is sufficient.

**Sample treatment:**
Filtration is performed over cleaned GF/C or GF/F glass fiber filters.

**Storage:**
Store in a cool place (4 °C). Filters are to be kept frozen at -18 °C. Addition of
chloroform or dichloromethane and BHT (bytylated hydroxy toluene) as anti-
oxidant is recommended.

**Analytical method(s):**
*Method #1.* The particulate phase is extracted in a Soxhlet apparatus using a
mixture of chloroform/methanol/water, while the filtered water phase is extracted with a mixture of chloroform and hydrochloric acid.
A preseparation step in phospholipids, glycolipids and neutral lipids involves
column chromatography under sequential elution with methanol, acetone and
chloroform.
After hydrolysis and trans-esterification the resulting fatty acid methylesters are
purified by thin layer chromatography (TLC). All steps should be carried out
under a nitrogen atmosphere.
Identification and quantification are performed by GC-MS.

**Remarks:**

**Method(s) references:**
Dawson & Liebezeit (1981); Saliot *et al*. (1988); Reemstma *et al*. (1990)

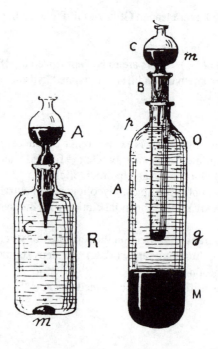

*The past:*
*Sounding instruments of Fol (l) and of Berget (r), both operating with mercury dropping into the lower chamber under pressure (From: Rouch, 1943)*

**A-7.24**

---

**Variable:**     **trace metals**

**Unit :**     **nmol/l**

**Compartment:**     **water**

**Introduction:**

Pollutants in the water column are present in the dissolved and in the particulate phase (SPM). The latter phase will be treated separately (A-7.36).

The group of trace metals usually analyzed for is rather small: Cd, Cu, Hg, Pb and Zn; while, of the non-metals As is often analyzed. But other elements are increasingly being analyzed for, either for toxicological reasons and not least, due to the availability of techniques that allow for multi-element detection (*e.g.* inductively coupled plasma mass spectrometry, ICP-MS; neutron activation analysis, NAA), thereby reducing the (relative) cost of analysis. These additional elements include Ag, Cr, Mo, Ni, Sb, Se, V, W and the rare earth elements (REE) (Merian, 1991; Johnson *et al.,* 1992). In addition, well defined chemical species are being identified (organo compounds of As, Hg, Pb and Sn). These compounds, which tend to be more toxic than the inorganic metal complexes, require specialist treatment and analysis, which is felt to be beyond the scope of this manual. Other 'chemical speciation' techniques result in chemical fractionation rather than speciation (Kramer & Duinker, 1984; Batley, 1989).

Most trace metals in sea and estuarine water are present in concentrations at µg/l level or below. Due to the availability of trace elements almost everywhere (dust), contamination control is of the utmost importance. Handling of open samples is carried out, preferably in a contamination free atmosphere (a clean room or, at least a clean bench), with all precautions taken (dust-free garments, plastic (polythene) gloves, etc.).

Many techniques are available for the analysis of trace elements in seawater. They often suffer from the salinity matrix (not the electrochemical techniques), and a separation step (which is also a concentration step) is included in the procedures. These separation/concentration steps usually involve the complexation of the trace metals (*e.g.* involving APDC/DDDC or HMA/HMDC mixtures of complexing agents) and subsequent extraction of the formed complex in an organic solvent (MIBK, freon, acetone). Back-extraction into an acidic solution (nitric acid) results in stable solutions, which can be analyzed by various instrumental methods (Danielsson *et al.,* 1982). Sorption columns (*e.g.* Chelex) have been used, but this is not in widespread use for estuarine studies.

The various atomic absorption methods are the most commonly applied electro-thermal AAS, with or without Zeeman correction, cold vapour AAS (for Hg); flame AAS is not too often applied due to the relatively high detection limit of the method), and electrochemical methods (differential pulse anodic stripping voltammetry (DPASV), chronopotentiometry and cathodic stripping voltammetry (CSV) (van den Berg *et al.,* 1991). These electrochemical methods allow analysis in seawater without pretreatment (ASV) or directly after complexation (CP, CSV) to be carried out. Depending upon the concentrations and elements, other techniques are used: optical ICP (ICP-AES), atomic fluorescence spectrometry (AFS, sensitive for Hg), radiochemical methods, total X-ray reflection (TXRF), and the forementioned ICP-MS and NAA (Stoeppler, 1991). Isotope dilution mass spectrometry (ID-MS), which is sometimes used to calibrate various methods is a reliable technique.

(Instrumental) analytical methods for trace elements compare reasonably well. Although large discrepancies may exist between laboratories (which is a cause for concern), no systematic differences have been observed between the analytical techniques proper.

The matrix in estuaries (salt content and relatively high organic matter content) poses special problems for the analysis of trace metals. Natural trace metal complexes appear to hamper the analysis, as it prevents total complexation (complexation/extraction techniques, underestimation) or a reduced calibration signal (electrochemical methods, overestimation). Destruction of the organic matter (*e.g.* involving a UV-digestion step) prior to analysis may be necessary. Certified reference materials for trace metals are on the market for sea- and estuarine water, which allow testing of the analytical procedures  to be performed (Cantillo, 1992).

**Sampling:**
Water may be sampled directly in the cleaned (acid washed) sample bottle. Bottles need to be pre-cleaned in steps using (1) washing with distilled water to remove dust; (2) leaching with reagent grade nitric acid (2:3); (3) leaching with Instra Analyzed grade nitric acid (1:12), each for $\geq 5$ days. After these cleaning steps, the bottles are filled with high purity water (Milli-Q or alike) and acidified with nitric acid (supra pure quality, 2 ml/l). This solution is kept in the bottles until use.

Due to the volatility of Hg, pre-cleaned glass bottles are used for this element. For sub-surface sampling, special trace metal free samplers which are lowered on a non-metal cable (polythene) must be used. When large volumes are needed, peristaltic pumping systems offer a good alternative (allowing for in-line filtration).

**Sample volume:**
For analysis, 0.5-1 litre is sufficient; polythene, polypropene or Teflon bottles are used.

**Sample treatment:**
In open ocean waters with a low SPM content, filtration is not often necessary and is a major risk of contamination. In estuaries, filtration needs to be applied when analyzing for dissolved trace metal concentrations. Samples are filtered over an acid pre-cleaned 0.45 µm membrane filter (cellulose acetate or cellulose

nitrate) or a 0.4 µm Nuclepore-type filter (polycarbonate). Filtration should be carried out in a 'class 100' clean room (clean bench) atmosphere.

After filtration the samples are acidified using 1-2 ml/l nitric acid (for Hg 5 ml/l) in each case of supra pure quality.

**Storage:**
When acidified to pH <2 samples can be stored at ambient temperatures without any problem. To minimize contamination, samples are packed in two polythene bags.

**Analytical method(s):**
*Method #1a.* Complexation/extraction, followed by (ET)AAS detection;
*Method #1b.* Complexation/extraction, followed by (ZET)AAS detection;
*Method #1c.* Complexation/extraction, followed by ICP-MS detection;
*Method #1d.* Complexation/extraction, followed by ICP-AES detection. Buffered samples are complexed involving (*e.g.*) APDC/DDDC or HMA/HMDC mixtures of complexing agents and subsequent extraction of the formed complex in an organic solvent (MIBK, freon, acetone). Back-extraction into an acidic solution (nitric acid) results in stable solutions which can be analyzed by various instrumental methods.
*Method #2.* Anodic stripping voltammetry (DPASV). After UV destruction, the acidified samples are analyzed directly in the voltammetric cell using a hanging mercury drop electrode or a mercury thin film electrode.
*Method #3.* Cathodic stripping voltammetry (CSV). A given trace element is complexed with an excess of a defined organic ligand. A potential is applied after which the complex is adsorbed to the surface of a stationary mercury drop electrode, which is measured (also called adsorptive differential pulse cathodic stripping voltammetry, ADPCSV).

**Remarks:**

**Method(s) references:**
Danielsson *et al.* (1982); Berman *et al.* (1983); Berman & Yeats (1987); Stoeppler (1991); van den Berg *et al.* (1991)

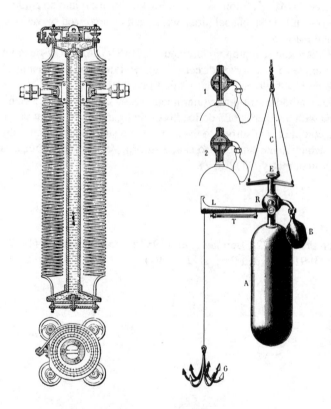

*The past:*
*Sounding instruments: the bathymeter of Siemens (1876) (From: Thoulet, 1890) and*
*the bathomètre of Regnard (From: Richard, 1907)*

**A-7.25**

---

**Variable:**          **polycyclic aromatic hydrocarbons**

                       **PAHs**

**Unit :**             µg/l

**Compartment:**    water

**Introduction:**
Polycyclic aromatic hydrocarbons (PAHs) are formed from both natural and
anthropogenic sources, the latter providing the major contribution. Large
quantities of PAHs are produced by the incomplete combustion and pyrolysis of
fossil fuels used as energy sources. Many of them have been found to be
mutagenic and/or carcinogenic (Neff, 1979).
There is an almost endless list of possible polycyclic aromatic hydrocarbons
(PAHs) in the environment, but, based on abundance and toxicity traditionally
only 6 PAHs were detected in environmental research (the 6 of Borneff). A better
understanding of the toxicological properties has led to a new list, the '16 of the
EPA', which has become a new standard (see A-7.37).
The lipophilic character of PAHs ensures that they bind strongly to particulates
and micelles, and it has been found that they are readily accumulated by
organisms, including plankton.
Since the PAHs are hydrophobic, a separation into a particulate and a dissolved
phase has been found to be problematic. If it is necessary to analyze the water
phase, then the best option seems to be to analyze the total water instead of
separate fractions. Since particulates are by far the most important phase, one
should consider analyzing the SPM only (see A-7.37).

**Sampling:**
As for all trace organic compounds, samples are collected in stainless steel or
glass sampling devices. For low concentrations, placing the collection directly in
the pre-cleaned (glass) sample bottle is preferred in order to avoid loss to the
sampler walls. A dark sample bottle (brown glass) minimises changes due to
photo-oxidation. Be aware that contamination problems of atmospheric origin
can occur: diesel oil seems to be present everywhere on a ship.
In order to prevent loss of the more volatile PAHs during storage, the sample
bottle should filled completely.

**Sample volume:**
Typical sampling volumes range from 1-5 litres, depending upon the expected
concentration range.

**Sample treatment:**
When filtration is to be applied (see remarks above), a (acetone and/or n-hexane) pre-cleaned glass fibre filter (GF/C or GF/F, 1.2 and 0.8 μm respectively) is used in a glass or stainless steel filter holder. An extraction thimble may be used instead. To minimise atmospheric interference (contamination) a closed in-line filtration system is preferred.

**Storage:**
Storage at 4 °C in the dark. Preferably, the samples should be analyzed within 24 hours after sampling.

**Analytical method(s):**
*Method #1a.* When analyzing the total amount of PAHs (particulate and dissolved) the total sample is submitted to a liquid/liquid extraction *e.g.* using n-hexane. The extracts are reduced in volume and analyzed using reversed phase high performance liquid chromatography (RP-HPLC) with fluorescence and UV-absorption detection. Quantitation is based on calibration using external standards.
*Method #1b.* If filtration is applied, the dissolved phase is submitted to liquid/liquid extraction, the particulate matter is extracted by sonication or by Soxhlet extraction using (*e.g.*) n-hexane. The combined extracts are reduced in volume and analyzed using reversed phase high performance liquid chromatography with fluorescence and UV-absorption detection. Quantitation is based on calibration using external standards.
*Method #1c.* Alternatively, the aqueous sample part may be submitted to solid-phase extraction in order to selectively remove the PAHs.
*Methods #2a,b,c.* Instead of reversed phase high performance liquid chromatography (RP-HPLC) with fluorescence and UV-absorption detection, capillary gas chromatography with mass-spectrometric detection (GC-MS) in the selected-ion monitoring (SIM) mode may be used. Identification of individual PAH is based on combined retention data, the specific ions detected and (optionally) the ion-abundance ratios.
*Methods #3a,b,c.* Before extraction, a mixture of $^{13}C_{12}$ labelled PAHs is added to the sample as an internal standard (isotope dilution).
The sample is pre-treated by one of the Methods (#1a, #1b or #1c) described above, and analyzed using capillary gas chromatography with mass-spectrometric detection in the selected-ion monitoring mode.
Identification of individual PAHs is based on a combination of retention data, the specific ions detected and ion-abundance ratios.
Quantitation is based on the comparison of responses with those of corresponding internal standards.

**Remarks:**

**Method(s) references:**
Neff (1979); Lee *et al.* (1981); Ehrhardt *et al.* (1991)

**A-7.26**

---

**Variable:**     **polychlorinated biphenyls**

                **PCBs**

**Unit :**         **ng/l**

**Compartment:**   **water**

**Introduction:**
In total, 209 PCBs exist, from the mono to the deca-chlorobiphenyls. Due to the complicated chemical names of PCBs, Ballschmitter & Zell (1980) proposed a numbering system, which has been adopted by the IUPAC.
The lipophilic character of PCBs makes them attach strongly to particulate matter and micelles, and it has been found that they are readily accumulated by orga-nisms, including plankton. For a description of the most important (abundance, toxicity) PCBs see A-7.38.
Since PCBs are hydrophobic, a separation into a particulate and a dissolved phase has been found to be problematic. Different methods of separation resulted in entirely different concentrations and relative congener distributions (Hermans *et al.*, 1992). If it is necessary to analyze the water phase, the best option seems to be to analyze the total water, instead of separate fractions. Since particulates are by far the most important phase, one should consider analyzing the SPM only (see A-7.38).

**Sampling:**
Trace organic compounds samples are collected in special stainless steel or glass sampling devices. Pumping systems, using Teflon tubing, have been applied successfully. Plastic devices (other than Teflon) are not to be used due to reasons of contamination. For low concentrations, collection directly into the pre-cleaned (glass) sample bottle or stainless steel tank is preferred, in order to avoid loss to the sampler walls. An dark sample bottle (brown glass, stainless steel) minimises changes due to photo-oxidation.

**Sample volume:**
Typical sampling volumes range from 1-5 litres, depending upon the expected concentration range.

**Sample treatment:**
When filtration is to be applied (see remarks above), a (n-hexane) pre-cleaned glass fibre filter (GF/C or GF/F, 1.2 and 0.8 µm respectively) are used in a stainless steel filter holder. Alternatively an extraction thimble may be used. Filter size for large volumes is typically 145 mm ø. To minimise atmospheric interference (contamination) a closed in-line filtration system is preferred.

**Storage:**
Storage at 4 °C in the dark. PCBs are not particularly susceptible to chemical degradation. In order to prevent loss of volatile PCBs, samples may be stored in a refrigerator. Sample flasks should preferably be completely filled.

**Analytical method(s):**
*Method #1a.* The aqueous sample part is submitted to liquid/liquid extraction, filtered particulate matter is extracted by sonication or by Soxhlet extraction using dichloromethane/n-hexane. The combined extracts are reduced in volume and cleaned by column chromatography using a suitable sorbent such as florisil, silica gel or aluminum oxide. The cleaned extract is concentrated and analyzed by capillary gas chromatography (GC) with electron capture detection (ECD). Identification of individual PCB is based on retention data, preferably using two GC columns with different polarity. Quantitation is based on calibration using external standards.
*Method #1b.* Alternatively, the aqueous fraction may be submitted to solid-phase extraction in order to selectively remove PCBs.
*Methods #2a,b.* Instead of capillary gas chromatography with electron capture detection capillary, gas chromatography with mass-spectrometric detection in the selected-ion monitoring mode may be used.
Identification of individual PCBs is based on combined retention data, the specific ions detected and (optionally) ion-abundance ratios.
*Methods #3a,b.* Before extraction, a mixture of $^{13}C_{12}$ labelled PCBs is added to the sample as an internal standard (isotope dilution).
The sample is treated by one of the Methods (#1a or #1b) described above, and analyzed using capillary gas chromatography with mass-spectrometric detection in the selected-ion monitoring mode.
Identification of individual PCBs is based on a combination of retention data, specific ions detected and ion-abundance ratios.
Quantitation is based on the comparison of responses with those of corresponding internal standards.

**Remarks:**

**Method(s) references:**
Reutergård 1980; UNEP (1984); Duinker & Hillebrand (1983); Duinker *et al.* (1988); Lazar *et al.* (1992)

**A-7.27**

---

**Variable:**        **suspended particulate matter**

                    **SPM**

**Unit :**           **mg/l**

**Compartment:**     **seston**


**Introduction:**
Suspended particulate matter (SPM, sometimes TSM, total suspended matter) or
seston consists of living particulate matter (plankton, etc), particulate organic
matter derived from dead organisms, condensates of organic matter, flocks, etc.,
and inorganic material such as sand, silt and clay. To a large extent biological
production and hydrodynamic conditions (currents, density; hence influencing
sedimentation, resuspension) determine the nature and abundance of the seston.
High variations in the total content and the relative composition, both in time and
place (along the estuary but also in the water column) will occur. SPM and
phytoplankton samples should be collected together from the same sample.
SPM is usually measured by back weighing of a filter after filtration. When a
large amount of suspended particulate matter is required, flow-through
centrifugation has been used. By (empirical) definition SPM consists of all
material that is retained by a 0.45 µm filter. The material collected by
centrifugation is not the same as that collected by filtration as the density of the
particles is of major importance; also the salt content cannot be eliminated.
Other methods for the estimation of the amount of suspended particulate matter
involve the various methods that are considered under the determination of
turbidity (A-7.4). (*In situ*) turbidity meters will give an estimate of the
distribution of particulate matter, but these methods have to be calibrated by
collection of samples as the measurements are dependent on the type of
particulate matter (size, form).

**Sampling:**
Sampling is best performed using relatively small water samplers (max. 1-2
litres), in which the total water content can be sampled. As the particulate matter has a tendency to be deposited, an overestimation may occur if only part of
the sampler content is collected. This will particularly occur if the filter is
mounted directly on the sampler because of the time necessary for filtration.
Shaking of the sampler is essential to keep the SPM in suspension and
homogeneously distributed. Horizontally mounted samplers, that enable
undisturbed passage when open, are considered the best option in high turbid
waters. Practical experience will be necessary to estimate the amount of water
that can pass a filter before it is clogged by the seston.

Collection directly in a sample bottle or by a pumping system may segregate the particles and give incorrect results.

For the collection of large volumes for flow-through centrifugation, the centrifuge may act as a pumping system. Otherwise an additional pump can be used to bring the water to the centrifuge.

**Sample volume:**

Usually 1 litre is sufficient to determine the amount SPM; in waters of low turbidity more water may be required (2-5 l). The samples are collected in a narrow mouth bottle of sufficient size.

For centrifugation volumes of 100 - 1000 litres are not uncommon. To be able to calculate the SPM content, the volume passing the centrifuge must be determined.

**Sample treatment:**

The filtration step for collection of the SPM over 0.45 μm filters (either membrane or glass fibre types) should be terminated by adding distilled water to remove the salt, taking care not to extract the organic matter. The amount of distilled water should preferably be in the order of 20 - 50 ml, but in high turbidity waters this amount may not pass the filter. At least a few ml should be applied in any case. Filters are air dried at 60 °C.

**Storage:**

Biological activity may change the SPM content. The filtration should prefera-bly be carried out directly after sampling. In between, the samples should be stored cool in the dark.

Dry filters are best stored in petri dishes of convenient size. While still wet, prevent the filter sticking to the glass, by leaving it on the rim of the petri dish for a while.

The SPM collected by centrifuge may be stored deep frozen (- 18 °C or better) for several months without major alteration.

**Analytical method(s):**

*Method #1.* Filters (0.45 μm, polycarbonate membrane, 0.4 μm Nuclepore, or glass fibre types, *e.g.* GF/F), are pre-weighed after drying in an desiccator. The sample is thoroughly homogenised and a (pre-)defined amount (depending on the estimated amount SPM, 0.1 - 1 litre for filters of 50 mm ø) of suspension is filtered under vacuum or under moderate pressure (nitrogen or filtered air, about 100 kPa). The filter is then washed with distilled water to remove salt. The fil-ter is air dried and stored. Before back-weighing the filters are placed in a desic-cator until the weight is constant.

*Method #2.* Flow-through centrifugation. For large amounts of SPM tens to hundreds of litres of water are pumped through a convenient flow-through centrifuge either from a storage tank or directly from the water column. The optimum flow rate in terms of retaining the particles should be based on test experiments. The flow rate should be checked regularly and the amount of water that passed the centrifuge should be determined.

*Method #3.* Calibrated turbidity. Transmission or scattering is measured in discrete samples or *in situ*. The method should be calibrated by filtering the SPM over a 0.45 μm filter (see Method #1). See also A-7.4.

**Remarks:**

**Method(s) references:**
Duinker *et al.* (1979); Reuter (1980); Bewers *et al.* (1985)

**A-7.28**

---

**Variable:**        **particle size**

**Unit :**           **n\*10³/ml (per size class)**

**Compartment:**     **seston, ( sediment )**


**Introduction:**
In estuarine environments large variations may occur in the size of (suspended) particulate matter. From the very small colloidal material that mainly exists in the low salinity parts, the size may extend to an excess of 400 µm due to aggregation of small particles to flocculates. These flocculates are difficult to sample, as they are extremely fragile and easily break up into smaller parts (Gibbs, 1982; Gibbs & Konwar, 1983; Eisma, 1986). For flock size determination the use of an *in situ* camera technique offers minimum disruption of the flocks (Eisma *et al.*, 1990). Within the scope of many projects the sampling of flocks will not be considered of prime importance, however. It is taken then for granted that in the determination of particle size the large particles are underestimated.
For the interpretation of transects not only the size of particles in the water column, but also of the very sediment surface would be helpful. The silt fraction of sediments (< 63 µm) can conveniently be analyzed by the methods described below.
Electronic counting and sizing of particles by conductometric methods (*e.g.* Coulter Counter) has become wide spread in marine environmental studies. Another technique involves laser diffraction (*e.g.* Malvern Autosizer).
Flow-cytometry is a relatively new technique. It allows the detection of particles in a continuous flow through a thin channel. The range of particles that can be determined is relatively large. Usually size and color (fluorescence) of the particles are determined. This allows the distinction between detritus and (different) plankton cells. At present the instrument should be considered as a specialist tool. The resulting data sets usually involve a large number of data points, depending on the instrument, that cover the spectrum in size classes (even up to 256). Often the storage of all data in a database, comprising all measured variables is not considered useful (although this may be obtained from the original source). For practical reasons, a reduction to a limited number of size classes is required. For example, the maximum number of size classes for reporting in the JEEP92 programme is 20.
The classes which are defined will be dependent on the instrument and technique, hence a standard cannot be put forward here. The dataset should include those data that make a reconstruction of the frequency/size plot possible and thus size class and number of particles. The volume can be calculated from the diameter, assuming a spherical shape.

**Sampling:**
Sampling is best performed using relatively small water samplers (1-2 litres), of which the total water content can be sampled. As the particulate matter has a tendency to be deposited, an overestimation may occur if only part of the sampler content is collected. Collection directly in a sample bottle or by a pumping system may segregate the particles and may give incorrect results.

**Sample volume:**
One litre is more than sufficient, and should be collected in a plastic or glass bottle.

**Sample treatment:**
None. The use of fixatives should be avoided. Dilution of samples to facilitate measurement is discouraged.

**Storage:**
Samples can not be stored and should be analyzed as soon as possible after collection (within 6 - 12 hours). Biological activity will change the size/density spectrum of the samples. Before analysis the samples should be stored cool (as close as possible to the *in situ* temperature) and in the dark.

**Analytical method(s):**
*Method #1.* Conductometric analysis (*e.g.* Coulter Counter). These techniques involves the suction of a given sample volume through one or more counting tubes with specific aperture; particles displace an amount of water proportional to their size and number, this is detected in the aperture by a change in current. To cover the entire bulk of the estuarine suspended particulate matter (seston), a size range of about 3 to 100 spheric equivalent diameter (SED) should be detected. Experience suggests that at least three counting tubes are necessary: 100 μm, 280 μm and 560 μm. Generally the following conditions are used:

| SED (μm) | tube considered | size classes | counting volume |
|---|---|---|---|
| 1.55 - 6.17 | 100 | 2 - 8 | 0.5 ml |
| 6.17 - 24.55 | 280 | 4 - 10 | 2 ml |
| 24.55 - 97.68 | 560 | 6 - 12 | 20 ml |

Testing of the performance of each tube in natural estuarine water conditions is advised. To adequately sample the largest particles (> 50 μm SED) counting of substantial volumes will be necessary (*e.g.* 20 ml). This can be achieved by counting on 'time' and calibrating the volume sampled.
Blocking of the orifice, especially of the smallest tube, at high particle concentrations can be avoided by reducing the counting volume to 0.05 ml.
*Method #2.* Laser diffraction. A light beam (laser) is directed towards the water sample. The particles scatter the light, and the angle of the scattered light and its

intensity are a measure of the size of the particle. The method requires no calibration. A size spectrum can thus be calculated. A laser light beam is directed towards the water sample either in a flow-through or in a discrete volume cell.

*Method #3.* Flow-cytometry. This technique measures light scatter and fluorescence of single, individual particles passing a laser focus at high speed in a water jet. Typically, up to 2,000 particles per second are analysed, with a practical analysis speed of a few minutes per sample. Forward light scatter is a measure of particle size and the side scatter is most sensitive for internal and external stucture of the particle. Measured fluorescence is proportional to the amount of natural cellular pigments, or artificially introduced fluorescing probes to label specific (constituents of) cells. New developments allow the sorting of particles into various classes.

**Remarks:**

**Method(s) references:**
Parsons *et al.* (1984); Demers (1991);  Individual instrument manuals

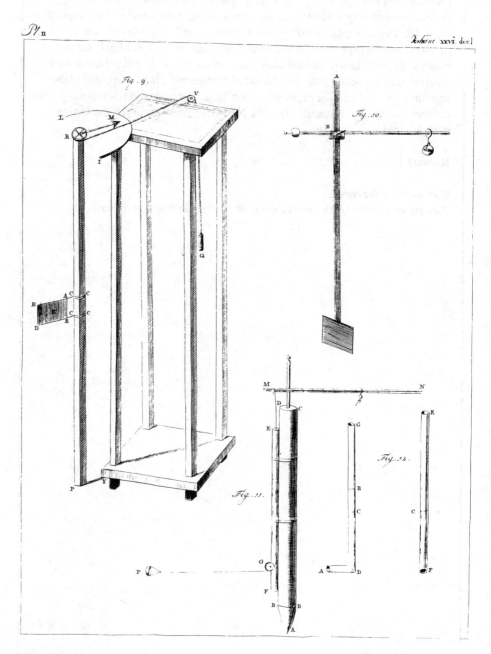

*The past:*
*Early attempts to measure currents from fixed locations (From: Brunings, 1789)*

**A-7.29**

---

**Variable:**         **chlorophyll** *a* **,** *b, c;* **other pigments**

                    **Chl a, b, c**

**Unit :**              µg/l

**Compartment:**    **seston**

**Introduction:**
Determination of particulate chlorophyll serves as an indicator of phytoplankton biomass.
The standard analytical methods for chlorophylls are based on spectrophoto-metry or fluorometry (Strickland & Parsons, 1968, 1972), and involve extraction in acetone/water or methanol/water mixtures. It should be understood that under the terminology 'chlorophyll' a series of pigments (chlorophylls, carotenoid pigments) and degradation products are covered. The spectrophotometric and fluorometric methods, despite their refinement, do not always provide the information needed on these various pigments found in natural waters.
In recent years, high-performance liquid chromatography (HPLC) methods have been developed that distinguish between the various pigment forms. These methods are more precise and may give additional insight into species composition and stage of plankton blooms through the pigment/degradation product spectrum. For these reasons the HPLC method is recommended. For low accuracy chlorophyll determinations, *i.e.* when ± 20% of the true chloro-phyll *a* concentration is acceptable, the standard spectrophotometric or fluorometric methods may be used.
*In situ* fluorometric detection of chlorophyll has become available. However, in high turbidity areas of estuaries they may fail to produce accurate results.
Several solvents are used to extract pigments from the particulate matter. Most common are acetone/water or ethanol/water (90%). The efficiency of ethanol (and methanol) is claimed to be better, especially when green algae or less readily extractable plankton species are present (Nusch, 1980). In an estuary, the salinity gradient and a distinct change in species composition may cause difficulties when using acetone as solvent. The ethanol/water mixture is there-fore preferred.
For most programmes primary importance should be given to the determination of chlorophyll *a*, *b* and *c*, and of phaeophytin.

**Sampling:**
Any sampling device can be used.

**Sample volume:**
Depending on the density of the phytoplankton, between 0.5 and 5 litres of estuarine water are sampled. Usually 1 litre is collected. Due to the patchiness of phytoplankton it is strongly suggested that the various physico-chemical analyses and the chlorophyll determination are made from the same water sample.

**Sample treatment:**
Prior to filtration the water samples must be pre-filtered through 200 μm mesh to remove zooplankton and large debris. Filtration over a 0.5 μm glass fibre (e.g. Whatmann GF/F) or a 0.45 μm membrane filter of 25 to 50 mm ø, should be performed as soon as possible after collection (< 4 hours). To do this a known volume of sample (for estuaries usually between 200 and 1000 ml for a 47 mm GF/F) is filtered, and some magnesium carbonate is added to the sample while filtering to prevent acidity of the filter.
Filters should either be immediately extracted or stored under controlled conditions.

**Storage:**
Biological activity (production, bacteria) will seriously interfere with the chlorophyll analysis. Therefore storage of the water sample is not possible. Until analysis the samples have to be stored cool and in the dark. Once collected on the filter, the filters have to be stored immediately deep frozen at - 20 °C or lower (up to 30 days); storage in liquid nitrogen (- 196 °C) is preferred.

**Analytical method(s):**
Filters should be completely ground in either ca 5 ml methanol containing 2% ammonium acetate, or 5 ml acetone/water 90% using either an ice- or $CO_2$-cooled tissue homogenizer under subdued light. The extract should be analyzed as soon as possible (within 6 hours) by HPLC, spectrometry or fluorometry.
*Method #1.* Reverse phase HPLC is used to separate the different components in the extract. Detection after separation is usually performed by a fluorescence ($\lambda_{ex} = 430 \pm 20$ nm, $\lambda_{em} = 665 \pm 20$ nm) and/or absorbence detector. In the chromatogramme the different peaks are identified and quantified. Identification of HPLC separated pigments should be based on coelution of standards and on spectral matching of the UV-visible spectra between 350-700 nm for the chlorophylls (and their break-down products).
*Method #2.* Spectrophotometric analysis of the extract in a spectrophotometer. The extinction is measured at 750, 664, 647, 630, 510 and 480 nm. The extinction at 750 nm is used to correct for the turbidity blank.
*Method #3.* Fluorometric analysis. Chlorophylls and degradation products are fluorescing compounds and can therefore be detected fluorimetrically by a (spectro)fluorimeter. The excitation at 431 nm is used together with an emission of 667 nm.
*In situ* fluorimeters have been designed for the continuous measurement of longitudinal or vertical profiles.

**Remarks:**

**Method(s) references:**
Holm Hansen *et al.* (1965); Lorenzen (1967); Jeffreys & Humphrey (1975);
Abayachi & Riley (1979); Nusch (1980); Aminot & Chaussepied (1983);
Gieskes & Kraay (1983); Mantoura & Llwellin (1983); Parsons *et al.* (1984);
Murray *et al.* (1986)

**A-7.30**

| Variable: | particulate organic carbon |
|---|---|
| | POC |
| Unit : | mg/kg |
| Compartment: | seston, sediment |

**Introduction:**

Seston consists of inorganic particles (silt, clay) and of organic matter, derived from *e.g.* living organisms (phyto- and zooplankton, bacteria), (pseudo)-faeces, and from their remains (lysed cell walls), or organic matter that is transported to the estuary from the basin slopes either by rivers or from the atmosphere. Flocculated or aggregated organic matter, for example humic substances, will also be part of the seston that will be characterized as particulate organic carbon. Organic colloids will usually not be trapped by the normally applied filtration techniques. POC represents, by its nature, a sum parameter that does not discriminate between the multitude of different organic components that can be present in the seston. POC is the main component of organic matter; the latter is commonly calculated by multiplying POC with a factor of two.

Rapid determination of POC became available with the persulphate oxidation method. More recently, dry combustion techniques were developed and these seems to be the universal method today. Care should be taken not to decompose the carbonates, that may be present on the filter, during combustion. This is usually performed by the inclusion of an acid treatment step prior to combustion. When this step is neglected, the combustion temperature should not exceed 500 °C. The spectrophotometric detection method, based on the wet oxidation of carbon by acid dichromate, usually gives higher results than those obtained by the persulphate method due to the oxidation of highly reduced compounds such as lipids.

Studies to evaluate the particulate organic matter content by microscopic observation give non reproducible results that are not comparable with the chemical methods.

The dry combustion method involving combustion of the filter in a muffle furnace and subsequent analysis based on the weight loss should not be used for estuarine work. This technique overestimates the organic matter content up to four times due to the included loss of crystal water in clay particles (Dankers & Laane, 1983).

For most estuarine work the collection will be by conventional samplers; sediment traps for the estimation of particulate fluxes will have only limited use under most estuarine conditions.

**Sampling:**
Most water-sampling techniques can be applied for the collection of POC.
Particulate matter may settle in the sampler. Care should be taken to collect a
representative sample from the sampler, by shaking shortly before collection,
or by collection of the entire volume of the sampler (see S-6.2). Centrifugation
gives different results from filtration, and is usually applied when large amounts
of seston have to be collected, *e.g.* for a characterization of specific organic
compounds.

**Sample volume:**
The collection of 1 litre of water will almost always be sufficient in estuaries.
If sub-samples are to be taken from the sample bottle for filtration, avoid
segregation of the sample.

**Sample treatment:**
The water sample is filtered over pre-treated (450 °C for 24 h) 0.8 µm glass fibre
filters (GF/F type). Silver filters are also used. Filtration should be performed as
soon as possible after collection, in order to avoid any decomposition of the
organic matter. A controlled vacuum (about 100 kPa) should be used. After the
sample has passed the filter, the salt is removed by rapidly passing through 20 ml
distilled water or ammonium formate isotonic solution (roughly equivalent to the
salinity of the estuarine sample). The latter method minimizes the rupture of
plankton cells. The filter should be manipulated by steel forceps to minimize
contamination.

**Storage:**
After collection rapid analysis is preferred. The filters can be stored folded in
pre-treated aluminium foil, at - 20 °C or lower in liquid nitrogen. Immediate
storage in a desiccator has also been reported as a suitable method.

**Analytical method(s):**
*Method #1.* Dry combustion. The glass fibre filters are folded in crucibles and
burned under oxygen in a furnace (500 - 800 °C). The evolved carbon dioxide
is then analyzed in a CHN or specific carbon analyzer.
*Method #2.* Persulphate oxidation. The organic matter on the filter is oxidized,
usually under pressure, using a solution of persulphate and phosphoric acid.
The evolving $CO_2$ is determined using a carbon dioxide sensitive infrared
detector.
*Method #3.* Spectrophotometric method. The organic matter is digested by "wet-
ashing" of the glass fibre filter with dichromate and concentrated sulphuric acid.
The decrease in extinction of the yellow dichromate solution at 440 nm is a
measure for the oxidizable carbon.

**Remarks:**

**Method(s) references:**
Menzel & Vaccaro (1964); Riley (1970); Copin-Montegut & Copin-Montegut (1973); Sharp (1974); Cauwet (1981); Aminot & Chaussepied (1983); Parsons *et al.* (1984); LeB. Williams (1985)

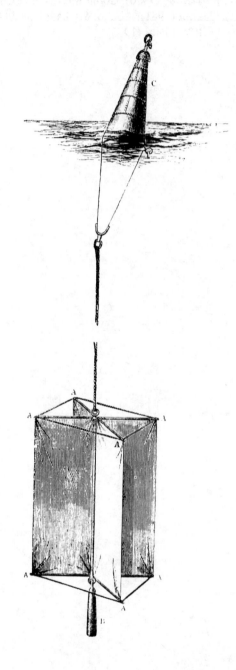

*The past:*
*The current drag used by the Challenger expedition (1873-76) (From: Challenger Report, 1885)*

**A-7.31**

| | |
|---|---|
| **Variable:** | **particulate organic nitrogen** |
| | **PON** |
| **Unit :** | **mg/kg** |
| **Compartment:** | **seston, sediment** |

**Introduction:**
As with POC, the particulate organic nitrogen will be a sum parameter. The sampling and sample treatment will be identical to the methods for POC. Both wet digestion and dry combustion methods have been developed, with either element specific detection or application of a CHN analyzer. The traditional method of distillation of the ammonia and titration with acid is not recommended because of the possible high blanks. The calculation of particulate organic nitrogen as the difference between total and dissolved nitrogen is also not recommended because of the low precision of the method.

**Sampling:**
Most water-sampling techniques can be applied for the collection of PON. Particulate matter may settle in the sampler. Care should be taken to collect a representative sample from the sampler, by shaking shortly before collection , or by collection of the entire volume of the sampler (see S-6.2). Centrifugation gives different results from filtration, and is usually used when large amounts of seston have to be collected.

**Sample volume:**
The collection of 1 litre of water will almost always be sufficient in estuaries. If sub-samples are to be taken from the sample bottle for filtration, avoid segregation of the sample.

**Sample treatment:**
The water sample is filtered over pre-treated (450 °C) 0.45 μm glass fibre filters (GF/F type). Silver filters are also used. Filtration should be performed as soon as possible after collection, in order to avoid any decomposition of the organic matter. A controlled vacuum (about 100 kPa) should be used. After the sample has passed the filter the salt is removed by rapidly passing through 20 ml of distilled water. The filter should be manipulated by steel forceps to minimize contamination.

**Storage:**
Rapid analysis after collection is preferred. The filters can be stored folded in pre-treated aluminium foil, at - 20 °C, or lower in liquid nitrogen.

**Analytical method(s):**
*Method #1.* Dry combustion. The glass fibre filters are folded in crucibles and burned in a furnace. A CHN analyzer is used for the detection of the nitrogen compounds evolved.
*Method #2.* Total Kjeldahl nitrogen. The sample is digested in aqueous sulphuric acid and the ammonia produced is determined in the neutralized digest by a colorimetric method.
*Method #3.* Persulphate digestion. The filter is digested with alkaline persulphate and the resulting nitrate determined as nitrite after reduction with cadmium.

**Remarks:**

**Method(s) references:**
Strickland & Parsons (1972); LeB. Williams (1985)

**A-7.32**

| Variable: | **particulate organic phosphorus** |
|---|---|
| | **POP** |
| **Unit:** | **mg/kg** |
| **Compartment:** | **seston, sediment** |

**Introduction:**
As with POC, the particulate organic phosphorus will be a sum parameter.
The sampling and sample treatment will be identical to the methods for POC.
Where filtration separates dissolved from particulate matter, oxidation serves to
distinguish between inorganic and organic phosphorus. Not all phosphorus is in
a form that is readily available to organisms. An example of the bio-available
phosphorus fraction from sediments has been estimated by Wiltshire (1991) for
the Elbe estuary. Particulate inorganic phosphorus can be determined by the
difference in reactive phosphorus from filtered and unfiltered samples. Analysis
of particulate material after oxidation will give a measure of its total phosphorus
content.
Wet digestion methods are usually applied, where the final analytical stage is
based on the determination of ortho-phosphate ions (A-7.11).

**Sampling:**
Most water-sampling techniques can be applied for the collection of POP.
Particulate matter may settle in the sampler. Care should be taken to collect a
representative sample, by shaking shortly before collection, or by collection of
the entire volume of the sampler (see S-6.2). Centrifugation gives different
results from filtration and is usually applied when large amounts of seston have
to be collected.

**Sample volume:**
The collection of 1 litre of water will almost always be sufficient in estuaries.
If sub-samples are to be taken from the sample bottle for filtration, avoid
segregation of the sample.

**Sample treatment:**
The water sample is filtered over pre-treated (acid leaching) 0.8 µm glass fibre
filters (GF/F type). Silver filters are also used. Filtration should be performed
as soon as possible after collection, in order to avoid any decomposition of the
organic matter. A controlled vacuum (about 100 kPa) should be used. After the
sample has passed the filter, the salt is removed by rapidly passing through
20 ml of distilled water. The filter should be manipulated by steel forceps to
minimize contamination.

**Storage:**
Rapid analysis after collection is preferred. The filters can be stored folded in pre-treated aluminium foil, at - 20 °C, or lower in liquid nitrogen.

**Analytical method(s):**
*Method #1.* Persulphate digestion. The filter is digested with alkaline persul-phate and the resulting orthophosphate determined by a spectrophotometric method (see A-7.11).
*Method #2.* Sulphuric acid/peroxide digestion. The filter with the residue is digested by heating with sulphuric acid and hydrogen peroxide ($H_2O_2$). The following determination of the ortho-phosphate can be determined either manually or by autoanalyzer method (cf. A-7.11).
*Method #3.* Perchloric acid digestion. The filter is treated with perchloric acid. After evaporation, the residue is heated to oxidize the organic matter and to liberate the phosphorus as inorganic phosphate which is determined by spectrophotometric analysis (A-7.11).

**Remarks:**

**Method(s) references:**
Strickland & Parsons (1972); Kattner & Brockmann (1980); Head (1985a); Yamada & Kayama (1987); Wiltshire (1991)

**A-7.33**

| | |
|---|---|
| **Variable:** | **clay content** |
| | **< 2 μm** |
| **Unit:** | **%** |
| **Compartment:** | **sediment** |

**Introduction:**
The terminology of sedimentary particle sizes is most frequently referred to the Udden-Wentworth Scale (Wentworth, 1922), which distinguishes between very coarse particles (such as boulders, cobbles and pebbles), sands, silt and clay. Each boundary between fractions is double the diameter of the previous one (see A-7.35). According to this scale clay is defined as material < 3.9 μm. Generally, however, clays are considered smaller than 2 μm, and are discrete particles, not aggregates (Gorsline, 1984).

The clay content of the sediment cannot be analyzed by (wet) sieving. The most frequently applied method involves the settling rates of the particles by either the "pipette" method or the Atterberg method. Differences between the methods were demonstrated to be substantial (Brennan, 1989). The principle of these analyses is that large particles in suspension in water fall faster than small ones, the fractions respectively deposition times can be calculated using the Stokes law. The procedure can be speeded up by the use of (batch) centrifugation.

**Sampling:**
The upper surface sediment layer gives the most recent information, although the physical and chemical information it contains will often be the net result of reworking by (bio)turbation. Many sediments are inhomogeneous because of this action. Organisms (and current or wave action) may 'sort' particles according to size and density, which will also result in the sorting of different (organic) constituents. To minimize the non-representativity of the sample, a number of sub-samples (5-10) may be collected that are pooled into one sample which is thoroughly homogenized. The top 0.5 cm is collected, taking care that the uppermost "liquid mud" layer is not washed away. Core samplers are most suitable for this purpose (see S-6.3).

**Sample volume:**
Depending on the diameter of the corer, between 1 - 50 cm$^3$ of sediment will be collected. For the determination of the clay content about 5 cm$^3$ is sufficient; for coarse sediments more material will be required and this may be collected by thorough mixing of several samples.

**Sample treatment:**
Both methods involve the dispersal of the finer particles using an anti-coagulant
such as ammonia or sodiumhexametaphosphate (Calgon method).

**Storage:**
Untreated samples may be stored at - 20 °C.

**Analytical method(s):**
*Method #1a.* Pipette method, peroxide treatment. For this method the sample is
first pre-treated with hydrogen peroxide ($H_2O_2$) to remove the organic coatings
that glue clay particles together, and with hydrochloric acid to remove
carbonates. The sediment is then brought into suspension at a preset temperature
(*e.g.* 20 °C). Three 20 ml pipette samples are collected from the water column
at a defined depth and after a defined time period of deposition. The clay content is measured by weighing.
*Method #1b.* Pipette method, no peroxide treatment. For this method the sample
is not pre-treated with hydrogen peroxide ($H_2O_2$). The further procedure is
equivalent to Method #1a.
*Method #2.* Atterberg method. About 2 - 5 g of sediment is brought into
suspension in an Atterberg tube, and left to rest at constant temperature (20 °C)
untill the end of the sedimentation time of the respective fraction. After this
defined sedimentation time (for the fraction < 2 µm: 23 h 22 min for a deposition length of 30 cm) the entire water column is collected. The particulate matter
is retrieved and determined by weighing. Other fractions may thus be collected
afterwards (*e.g.* < 6.3 µm after 2 h 21 min sedimentation time, and < 20 µm after
14 min sedimentation time; 30 cm deposition path).
*Method #3.* Conductometric analysis (*e.g.* Coulter Counter). The sediment is
brought into suspension, and analyzed by conductometric particle sizer (see A-
7.28).
*Method #4.* Laser diffraction (*e.g.* Malvern Autosizer). The sediment is treated
with peroxide to remove organic coatings, and with hydrochloric acid to remove
carbonates. The sediment is brought into suspension. A laser light beam is
directed towards the water sample in a stirred cell. The suspended particles in
he sample scatter the light, the angle of the scattered light and its intensity are
measured and interpreted in terms of size of the particle. From the size distribution spectrum the fraction < 2 µm can thus be calculated.

**Remarks:**

**Method(s) references:**
Müller (1964); Leeder (1982); Buchanan (1984); Instrument specific manuals

| Variable: | silt content |
| --- | --- |
| | **< 63 μm** |
| **Unit:** | **%** |
| **Compartment:** | **sediment** |

**Introduction:**
Between the arbitrarily selected boundaries of clay and very fine sand (according to the Udden-Wentworth Scale) is the silt fraction. It is assumed that this fraction, together with the smaller clay fraction, can be transported in suspension under "normal" conditions. Of course larger material can be transported by strong currents or turbulence.
The true silt content is the difference between the fraction < 63 μm (silt +clay) and the fraction < 2 μm (clay). As the determination of the clay content is rather laborious, often the silt+clay fraction is determined by wet sieving.
For most estuarine purposes the fraction < 63 μm is a suitable description of the very fine sediments. For practical reasons this method will therefore have preference over the determination of the true silt content.
Coagulated particles, *e.g.* originating from pseudo-faeces, may interfere with the analysis. Treatment with an oxidising medium (peroxide) will break the organic bonds between the clay/silt particles.
Although, for example, laser diffraction is a suitable technique for the determination of the particle size distribution, the sieving method is considered to be more simple to perform and is therefore preferred.

**Sampling:**
The surface sediment layer gives the most recent information, although the physical and chemical information it contains will often be the net result of reworking by (bio)turbation. Many sediments are inhomogeneous because of this action. Organisms (and current or wave action) may 'sort' particles according to size and density, which will also result in the sorting of different (organic) constituents. To minimize the non-representativity of the sample, a number of sub-samples (5-10) may be collected that are pooled into one sample which is thoroughly homogenized.The top 0.5 cm is collected, taking care that the uppermost "liquid mud" layer is not washed away. Core samplers are most suitable for this purpose (see S-6.3).

**Sample volume:**
Depending on the diameter of the corer, between 1 - 50 cm$^3$ of sediment has to be collected. For the determination of the silt content about 5 cm$^3$ is sufficient;

for coarse sediments more material will be required, when necessary these may be composed of separate samples which have been combined and homogenized.

**Sample treatment:**
Peroxide is sometimes used to decompose the organic coating that may glue small particles together, as mentioned under the methods of clay analysis.

**Storage:**
The untreated sample may be stored at - 20 °C.

**Analytical method(s):**
*Method #1.* Wet sieving. Depending on the estimated amount of fines, 10 - 50 cm$^3$ of sediment is gently washed through a 63 µm sieve using distilled water. The fraction < 63 µm is collected from the water that passes through the sieve. The coarse and fine fraction the material is dried at 95 °C. The weight is determined and the relative amount calculated [1]).
*Method #2.* Conductometric analysis (*e.g.* Coulter Counter). The sediment is brought into suspension, and analyzed by conductometric particle sizer (see A-7.28), which reveals also the particle size spectrum.
*Method #3.* Laser diffraction (*e.g.* Malvern Autosizer). The sediment is brought into suspension. A laser light beam is directed towards the water sample in a stirred cell. The suspended particles in the sample scatter the light, the angle of the scattered light and its intensity are measured and interpreted in terms of size of the particle. From the size distribution spectrum the fraction < 63 µm can thus be calculated.

**Remarks:**
[1])   The residue on the sieve can conveniently be used for the determination of the grain size distribution of the coarser material (A-7.35).

**Method(s) references:**
Leeder (1982); Buchanan (1984); Instrument specific manuals

**A-7.35**

| | |
|---|---|
| **Variable:** | **grain-size distribution** |
| **Unit:** | **%** |
| **Compartment:** | **sediment** |

**Introduction:**
The Udden-Wentworth Scale for the classification of sediments defines a series
of classes and sub-classes of sediment types.
The most important size-classes for work in estuaries are:

| *Name:* | *size range (μm)* |
|---|---|
| clay | < 2 |
| silt | 2 - 63 |
| very fine sand | 63 - 125 |
| fine sand | 125 - 250 |
| medium sand | 250 - 500 |
| coarse sand | 500 - 1000 |
| very coarse sand | 1000 - 2000 |

For estuarine work involving ecological/biological interactions, the coarser
classes belong to sediment types that are not particularly interesting. Therefore
the description usually ends with the fraction > 2 mm.
Dry and wet sieving are both popular. The disadvantage of dry sieving is the
possibility that clayish sediments when dried, tend to form lumps that consist of
fine material, thus disturbing the analysis. This problem is solved, when the
> 63 μm fractions are determined after the separation of the silt+clay fractions
by wet sieving.

**Sampling:**
The surface layer of the sediment gives the most recent information, although
the physical and chemical information it contains will often be the net result of
reworking by (bio)turbation. Many sediments are inhomogeneous because of this
action. Organisms (and current or wave action) may 'sort' particles according to
size and density, which will also result in the sorting of different (organic)
constituents. To minimize the non-representativity of the sample, a number of
sub-samples (5-10) may be collected that are pooled into one sample which is
thoroughly homogenized. The top 0.5 cm is collected. Core samplers are most
suitable for this purpose, although grab samplers may be used (see S-6.3).

**Sample volume:**
Depending on the diameter of the corer, about 1 - 50 cm$^3$ of sediment will be
collected. For the determination of the grain size distribution about 100 cm$^3$ is

sufficient. Sufficient material may be collected by thorough mixing of several samples.

**Sample treatment:**
No special treatment is required. Ensure that the sample is prevented from drying out.

**Storage:**
Samples may be stored at - 20 °C until analysis.

**Analytical method(s):**
*Method #1a.* Dry sieving[1]). The material left on the 63 μm sieve is used for the calculation of the < 63 μm fraction (see A-7.33) and is quantitatively transported to a petri-dish. About 100 g of sediment is needed for the analysis. After drying (at 95° C), the size-classes are determined by sieving over a set of nested sieves of appropriate mesh size under vibration. The sieving step should be standardized at 15 min. The separate fractions are measured by weighing and the percentages calculated.
*Method #1b.* Wet sieving. The (wet) sediment is passed over a set of nested sieves of appropriate mesh size under vibration and applying a gentle water flow. The separated fractions are dried, measured by weighing and the percentages calculated.

**Remarks:**
[1]) This dry-sieving method can only be applied when the silt+clay fractions have been removed.

**Method(s) references:**
Leeder (1982); Buchanan (1984); Giere *et al*. (1988)

**A-7.36**

**Variable:** **particulate trace metals**

**Unit:** **mg/kg**

**Compartment:** **seston, sediment**

**Introduction:**

Since many trace metals are sorbed onto (suspended) particulate matter, after deposition they may accumulate in sediments. In vertical profiles obtained from slices from sediment cores, the pollution history can often be constructed. As in water, there is an increasing interest in other than the standard elements (Zn, Cd, Pb, Zn, Hg), including Ag, Cr, Mo, Ni, Sb, Se, V, W and the rare earth elements (REE) (Merian, 1991; Johnson *et al.*, 1992). In addition, well defined chemical species are being identified (organo compounds of As, Hg, Pb and Sn).

Due to the natural variability of sediments (grain size, organic matter content, carbonates), the distribution of pollutants (and their availability to organisms) is not only determined by the hydrodynamics of the system, but also to a large extent by the characteristics of the sediment. When evaluating results of (whole) sediment analyses it will become clear that sediment grain size in particular plays an important role in the distribution of pollutants and other compounds: generally, the coarser the sediment the lower the pollutant content (dilution). Therefore, to be able to compare and characterize different sediments for their pollutant load, even when collected at relatively short distances, the inhomogeneous nature of the sediments makes some form of standardisation necessary. Two principal routes of normalization procedures have been followed. They include either the separation of a specific grain size fraction (granulometric normalization) and the analysis of contaminants therein, or the normalisation of whole sediments using reference (or conservative) elements.

The fraction <63 µm (silt + clay), obtained by wet sieving has become a frequently used standard 'fine' fraction, but also the <20 µm fraction has been used. The geochemical normalization is based on the observation that fine particulates are rich in clay minerals, iron and manganese oxy-hydroxides and organic matter, and that the coarser fractions are chemically inert. Specific elements like Al, Li or Sc can be used as tracers to normalize for grain size effects.

These two approaches lead to differences in the analytical methodology: total sediment or fine fraction. Due to their different entities, they are used as different methods here.

In addition, no agreement has been reached on the method of digestion. For a true 'total' trace metal content the digestion acid mixture requires the addition of hydrofluoric acid (HF). Only in this case are sediment matrices truly dissolved, which is essential to estimate the total amount of trace elements and of the normalizing elements. The HF is applied together with one or more different acids.

However, since the analysis involving HF may result in a more difficult matrix to analyze, 'partial' destruction methods are often applied, *e.g.* using *aqua regia*. The problem of 'partial' is that many acids or combinations of acids are possible, that the characteristics of the sediments are not constant and 'partial' is, thus, poorly defined (Loring & Rantala, 1990).

Destruction usually takes place in teflon bombs at elevated temperatures and pressure (oven or microwave), but other (open) methods are used (Bock, 1979). Once destruction is performed, many analytical techniques are available for the analysis of trace elements in sediments (Stoeppler, 1991). The methods given for the analysis of trace metals in seawater (A-7.24) may, in principle, be applied to dissolved particulates. Some methods, like NAA do not require a destruction step.

(Instrumental) analytical methods for trace elements compare reasonably well. Although large discrepancies may exist between laboratories, no systematic differences have been observed between the techniques proper.

The results are expressed on the basis of dry weight (DW).

For trace metals in sediments, certified reference materials (based on total digestion) have been made available, allowing for the testing of analytical procedures (Cantillo, 1992).

**Sampling:**
Sediment samples for trace metal analysis need to be collected contamination free. Thus, the sampler should contain no metal parts. When this is unavoidable, the sample should be collected from the central part of the bulk sediment, which has not been in contact with the sampling gear (See S-6.3).

The sample is collected in acid precleaned, polythene, wide-mouth bottles.

**Sample volume:**
For the analysis of total sediment or a grain size fraction 0.2-1 g is sufficient; a larger sample (10-20 g) that is homogenized before sub-sampling is preferred, however.

**Sample treatment:**
For total sediment analysis no treatment is necessary.

To collect a grain-size fraction, the sample is wet sieved over a nylon sieve of 63 μm mesh size (or 20 μm).

**Storage:**
Samples should be kept cool (4 °C). Unless analysis can start within a day after collection the samples are stored deep frozen (-18 °C).

**Analytical method(s):**
*Method #1.* Total digestion, total sediment, followed by instrumental analysis.
*Method #2.* Total digestion, <63 µm fraction, followed by instrumental analysis.
*Method #3.* Partial digestion, total sediment, followed by instrumental analysis.
*Method #4.* Partial digestion, <63 µm fraction, followed by instrumental analysis.
Total digestion involves total dissolution of the crystal latices of the sediment
matrix, using a mixture of HF together with (*e.g.*) aqua regia ($HNO_3$-HCl), nitric
acid or nitric/perchloric acid.
Partial digestion involves a destruction step using one or more of these acids
(*e.g.* HCl, $HNO_3$, $HClO_4$), thus no HF is applied here.

**Remarks:**
The use of perchloric acid requires a special safety fume hood.

**Method(s) references:**
Baudo (1990); Mudroch & Bourbonniere (1991); Merian (1991); Stoeppler
(1991); EPA (1992)

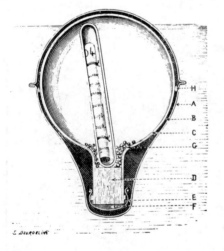

*The past:*
*Drift 'bottles', used for surface current measurements by Albert 1er de Monaco*
*(From: Albert 1er, 1932)*

**A-7.37**

**Variable:**         **particulate polycyclic aromatic hydrocarbons**

                         **PAHs**

**Unit:**                   µg/kg

**Compartment:**     seston, sediment

**Introduction:**
Polycyclic aromatic hydrocarbons (PAHs) are formed from both natural and anthropogenic sources, the latter providing the major contribution. Large quantities of PAHs are produced by the incomplete combustion and pyrolysis of fossil fuels used as energy sources. Many of them have been found to be mutagenic and/or carcinogenic (Neff, 1979).
Traditionally only 6 PAHs were detected in environmental research (the 6 of Borneff), but at present a wider spectrum is analyzed, and the '16 of the EPA' has become a standard, involving naphtalene, acenaphthylene, acenaphthene, fluore-ne, phenanthrene, anthracene, fluoranthene, pyrene, benzo[$a$]anthracene, chrysene, benzo[$b$]fluoranthene, benzo[$k$]fluoranthene, benzo[$a$]pyrene, benzo[$ghi$]perylene, dibenzo[$ah$]anthracene and indeno[1.2.3-$cd$]pyrene.
To allow a comparison of the toxicological effect of PAHs to be made, toxic equivalent factors (TEFs) have been determined from a number of them (Nisbet & LaGoy, 1992).
For the determination of PAHs in sediments, usually a known quantity of sample is extracted by solvents, often subsequently (*e.g.* acetone, n-hexane and water), after which the extract is concentrated before instrumental analysis.
Certified mixtures of PAHs as a standard reference material are available to check on the analytical performance (Cantillo, 1992).

**Sampling:**
SPM may be collected on a (acetone or n-hexane) pre-cleaned glass fibre filter (GF/C or GF/F, 1.2 and 0.8 µm respectively), or by the (continuous) centrifugation method.
Cleaning prior to collection of all parts involved in the sampling operation is essential to avoid contamination.

**Sample volume:**
10-50 g, depending upon the concentration levels expected.

**Sample treatment:**
None required.

**Storage:**
Storage at 4 °C in the dark. Preferably, the samples should be analyzed within 24 hours after sampling.

**Analytical method(s):**
*Method #1.* The sample is homogenized and submitted to solvent extraction alternatively using acetone and n-hexane or petroleum ether.
The acetone is removed by back-extraction with water.
The extract is reduced in volume and cleaned by column chromatography using a suitable sorbent such as florisil, silica gel or aluminum oxide.
The cleaned extract is concentrated and analyzed using reversed phase high performance liquid chromatography (RP-HPLC) with fluorescence and UV absorption detection. Quantitation is based on calibration using external standards.
*Methods #2* Instead of reversed phase high performance liquid chromatography with fluorescence and UV absorption detection, capillary gas chromatography with mass-spectrometric (GC/MS) detection in the selected-ion monitoring (SIM) mode may be used. Identification of individual PAHs is based on combined retention data, the specific ions detected and (optionally) ion-abundance ratios.
*Methods #3.* Before extraction, a mixture of $^{13}C_{12}$ labelled PAHs is added to the sample as an internal standard (isotope dilution).
The sample is treated by one of the Methods #1a-c, described above, and analyzed using capillary gas-chromatography with mass-spectrometric detection in the selected-ion monitoring mode.
Identification of individual PAHs is based on a combination of retention data, the specific ions detected and ion-abundance ratios.
Quantitation is based on the comparison of responses with those of corresponding internal standards.

**Remarks:**

**Method(s) references:**
Neff (1979); Lee *et al.* (1981); Ehrhardt *et al.* (1991)

**A-7.38**

**Variable:**          **particulate polychlorobiphenyls**

'        **PCBs**

**Unit:**              **ng/kg**

**Compartment:**    **seston, sediment**

**Introduction:**
In total 209 PCBs exist, from the mono to the deca-chlorobiphenyls. Due to the complicated chemical names of PCBs, Ballschmitter & Zell (1980) proposed a numbering system, which has been adopted by the IUPAC.
PCBs vary considerably in toxicity. Most toxic congeners are considered to be those that have a flat structure, the so-called co-planar congeners. In the natural environment these co-planar PCBs (IUPAC no. 15, 37, 77, 81, 126, 169) are not usually the most abundant, and it is common practice to analyze other, more abundant congeners as well.
The EPA preliminary recommendations for congener specific PCB analysis group the PCBs (in dredged material) into separate classifications (Clarke *et al.*, 1989):
Class 1: highest priority congeners, which are most likely to contribute to any adverse biological effects; subdivided into two classifications:
Class 1A: MC-type inducers (present at low concentrations) (IUPAC no. 77, 126, 169) and Class 1B: mixed-type inducers, commonly found in environmental samples (IUPAC no. 105, 118, 128, 138, 156, 170);
Class 2: PB-type MFO inducers, commonly found in environmental samples (IUPAC no. 87, 99, 101, 153, 180, 183, 194);
Class 3: moderate priority in assessing the potential of a PCB contaminated sediment (IUPAC no. 18, 44, 49, 52, 70, 74, 151, 177, 187, 201);
Classification 4: low priority (37, 81, 114, 119, 123, 157, 158, 167, 168; 189).
Not all PCBs can reliably be detected at present, due to analytical constraints (very low concentrations, congeners appearing at the same location in the chromatogram, etc.).
Apart from abundance and toxicity other criteria may be included, such as the existence of toxicity equivalent factors (TEFs), which normalise the toxicity relative to Dioxin (IUPAC nos. 77, 126 and 169; 105, 114, 118, 123, 156, 157) (Safe, 1990), or conformation to existing local or internationally accepted selections of series PCB congeners.
Typical indicator PCBs, often included in routine monitoring studies include: IUPAC nos. 28, 52, 101, 118, 138, 153 and 180. For ICES intercomparison exercises the nos. 105 and 156 are added to this list.

**Sampling:**
SPM samples are collected by filtration over a glass fibre filter (GF/C or GF/F). Sediment is collected by coring or by grab sampler (see S-6.3). Special care is taken not to contaminate the sample. No plastic sampler parts or tools should be used. The central parts of the sediment, which have not been in contact with the sampler walls, are preferred.

**Sample volume:**
10-50 g, depending upon the concentration levels expected.

**Sample treatment:**
None required.

**Storage:**
PCBs are not particularly susceptible to chemical degradation. In order to prevent loss of volatile PCBs, samples may be stored in a refrigerator.
Sample flasks should preferably be completely filled.

**Analytical method(s):**
*Method #1a.* The sample is homogenized and submitted to solvent extraction, alternatively using acetone and n-hexane or petroleum ether. The acetone is removed by back-extraction with water. The extract is reduced in volume and cleaned by column chromatography using a suitable sorbent such as florisil, silica gel or aluminum oxide. The cleaned extract is concentrated and analyzed by capillary gas chromatography (GC) with electron capture detection (ECD). Identification of individual PCB is based on retention data, preferably using two GC columns with different polarity. Quantitation is based on calibration using external standards.
*Methods #1b,c.* Alternatively, the sample may be submitted to extraction with acetone/n-hexane using sonication or Soxhlet extraction.
*Methods #2a,b.* Instead of capillary gas chromatography with electron capture detection capillary gas chromatography with mass-spectrometric detection in the selected-ion monitoring mode may be used. Identification of individual PCBs is based on combined retention data, specific ions detected and (optionally) ion-abundance ratios.
*Methods #3a,b.* Before extraction, a mixture of $^{13}C_{12}$ labelled PCBs is added to the sample as an internal standard (isotope dilution). The sample is treated by one of the Methods (1a-c) described above, and analyzed using capillary, gas chromatography with mass-spectrometric detection in the selected-ion monitoring mode. Identification of individual PCBs is based on a combination of retention data, the specific ions detected and ion-abundance ratios. Quantitation is based on the comparison of responses with those of corresponding internal standards.

**Remarks:**

**Method(s) references:**
Erickson (1986); Waid (1986, 1987)

**A-7.39**

| Parameter: | **Bacteria** |
| --- | --- |
| | **bacteria numbers (and biomass)** |
| Unit: | **n\*10$^6$/ml** |
| Compartment: | **water** |

**Introduction:**
Total bacterial numbers in sea- and estuarine water can be estimated from direct counts under a fluorescence microscope. Other methods involve culturing techniques in specific media, which includes the most probable number (MPN) technique. These methods are rather selective for bacteria that grow faster in a defined medium. The first method is considered the simplest and most widely applied method for estimating total bacteria numbers.
Bacterial biomass can be estimated from the calculation of bacterial size (in $\mu m^3$) measured under a microscope (epifluorescene microscopy). After determining the bacterial biovolumes, the biomass is calculated using the conversion factor of $5.6 \times 10^{-13}$ g C/$\mu m^3$ (Bratback, 1985).

**Sampling:**
Samples used for MPN etc. should be collected using a sterile water sampler (*e.g.* prewashed with 70% ethanol), or a special sampler that can be heat sterilized. To avoid the problem of a patchy distribution, *e.g.* due to the presence of bacteria in particulate aggregates, a number of samples (ca. 20) should be collected and these are pooled to get the final sample that is used for counting (see A-7.40).

**Sample volume:**
About 10 ml sample is sufficient for the counting.

**Sample treatment:**
About 10 ml of the estuarine water is transferred to a (sterile) screw-cap vial, and 0.5 ml of a 40% formaldehyde (or glutaraldehyde) solution is added to preserve the sample. The preservation step may be ignored if the sample is to be analyzed immediately.

**Storage:**
Preserved samples should be stored in a cool dark place (10 °C) and analyzed within 2 weeks.

**Analytical method(s):**

*Method #1.* Fluorescence microscopy. Special black filters (2.5 cm ø, 0.2 μm
pore size) may be obtained, or filters should be stained black with an Irgalan
black solution. After rinsing the pipette with the sample to be analyzed, 2 ml
sample is introduced in the filter holder, 0.2 ml Acridine orange or DAPI
solution is added to stain the bacteria and after 2 min incubation a moderate
vacuum (50 kPa) is applied for filtration. The dry filter is placed on a slide,
covered with microscopy oil, and counted within 24 h; at least 10 fields should
be counted with about 20 to 40 bacteria per field. To reach this, dilution of the
sample may be required.

**Remarks:**

Formalin is a toxic substance. Please refer to the safe handling procedures for
chemicals.

**Method(s) references:**

Vollenweider (1969); Hobbie *et al.* (1977); Parsons *et al.* (1984)

**A-7.40**

---

| Variable: | **Bacteria** |
| | |
| | **bacterial production** |
| | |
| Unit: | **mg C/l.h** |
| | |
| Compartment: | **water** |

**Introduction:**
Production of heterotrophic bacteria in marine, estuarine and freshwater environments can be estimated by measuring the incorporation of radio-labelled (tritium, $^3$H) thymidine ([methyl-$^3$H]thymidine) into DNA (tritiated thymidine incorporation, TTI). Another method, which is far more time consuming involves the determination of the frequency of dividing cells. This method involves no radiotracer work.

**Sampling:**
Most water sampling techniques can be used, but collection using a sterile water sampler (*e.g.* prewashed with 70% ethanol), or special sampler that can be heat sterilized is preferred. Care must be taken to collect a representative sample, especially in high turbidity zones. The pooling of a number of separate samples (20) is recommended.

**Sample volume:**
For most estuaries and coastal waters 50 ml samples will be sufficient for a single series of analyses.

**Sample treatment:**
As the turn-over of bacteria is high (reproduction rate about 20 min$^{-1}$), rapid action is necessary to estimate the production rate.
Of a set of six sub-samples of 5 ml each, three blanks are treated with formaldehyde to give a final concentration of 1.5 - 2 %. All sub-samples are spiked with tritiated thymidine and incubated for 30 - 45 min at *in situ* (field) temperature, in the dark. After incubation, the three sub-samples that were not preserved, are spiked with formaldehyde to stop bacterial growth.

**Storage:**
Storage of the water sample for later analysis is not possible, immediate action is required. After incorporation of the thymidine the vials can be stored on ice until filtration. When plastic scintillation vials are used the radioactivity should be measured within 24 h after addition of the scintillation fluid.

**Analytical method(s):**
*Method #1.* TTI. After the actual incorporation step (see above), the samples are filtered over a 0.2 µm membrane filter, which has been wetted with 1 ml 5 mM unlabelled thymidine. Vacuum filtration under mild vacuum (56 kPa). After washing with trichloroacetic acid (5 % (w/v) the filters are transferred to scintillation vials, scintillation fluid is added and the radioactivity measured by scintillation counting. From the incorporated activity the bacterial production can be calculated.
*Method #2.* Growth rate. This method involves the determination of the frequency of dividing cells for the determination of bacterial production.
*Method #3.* Dialysis. A method without the use of radio-labelled tracers uses dialysis membranes. The sample is filtered over a filter of 3-5 µm to get rid of particulate matter and bacteria consuming organisms (bacteriovores). The filtrate is incubated *in situ* in a dialysis bag. The (total) numbers of the bacteria are estimated in the sample by epifluorescence microscopy at the start of the experiment ($t_0$), and after a few hours ($t_1, t_2, t_3, ...$), from which the number of bacteria produced per volume and unit time can be calculated. This value is compared with a non-filtered sample which is incubated and analysed similarly to estimate the effect of predation by bacterial consumers (*e.g.* ciliates, flagellates).

**Remarks:**
As the first method involves radioactive material the proper precautions should be maintained, and the national safety regulations are to be followed.
Formalin is a toxic substance. Please refer to the safe handling procedures for chemicals.

**Method(s) references:**
Sorokin (1972); Hagström *et al.* (1979); Fuhrman & Azam (1982)

**A-7.41**

---

| **Variable:** | **Phytoplankton** |
| | **species abundance** |
| **Unit:** | **n/ml** |
| **Compartment:** | **water** |

**Introduction:**

Phytoplankton is present in estuarine waters together with detrital and inorganic particles. The difficulty in analyzing phytoplankton is to separate living cells from dead cells, debris and other amorphous particles. Automatic methods are difficult to use and direct examination under a microscope is therefore the most likely method. From the preserved sample the micro-zooplankton should be analyzed.

As a first step, the analysis can be limited to the identification of dominant phytoplankton types. The determination of the ratio of phytoplankton cells to other particles is often not possible in turbid estuarine waters. When the flagellates, which are often the dominant species, are to be identified to the species level (which is not always necessary), it is essential that these fragile organisms are analyzed immediately in the non-preserved samples. This is also the case when heterotrophic and autotrophic (chlorophyll containing) flagellates are to be distinguished (see below).

A problem in estuarine environments, especially in the upper parts, is that sometimes large quantities of freshwater plankton species are washed into the estuary by the river. To identify these species a (non-marine) specialist is required. Even then, as the alter in salinity may alter their characteristics, identification might be problematic.

The method described here assumes pre-concentration of the sample; this is not necessary when high densities are found.

**Sampling:**

Water samples are collected with any type of water sampler or pumping system, or are collected directly in the sample bottle. For optimal representativity see the sampling procedures for SPM (A-7.27).

**Sample volume:**

A sample volume up to one litre is collected.

**Sample treatment:**

No preservation is needed for immediate analysis of flagellates. In other cases preservation of the samples is essential. Phytoplankton and planktonic protozoa for microscopic analysis can be preserved with a small amount of buffered formalin (enough to bring the sample to 1 % formalin) or with acidic Lugol's solution (made by addition of 50 g iodine and 100 ml acetic acid to a solution of 100 g KI in 1 l distilled water). Other methods of preservation involve the use of neutralized formalin or hexamethylenetetramine.

When samples are preserved in glutaraldehyde (1.5-3% final concentration) or in a mixture of both preservatives, flagellates stay intact and allow fluorescence microscopic detection and distinguishing between heterotrophic and autotrophic (micro)flagellates.

When picoplankton is present (though this will not often be the case in estuarine environments) discrimination between eukariotes and prokariotes (cyanobacteria, prochlorophytes) has to be performed by epifluorescence microscopy. This can only be carried out on live samples or on samples fixed with glutaraldehyde. Analysis in this case should be within a few days as the fluorescing properties may deteriorate with time.

At low densities preconcentration will be necessary. After preservation of the sample, the bottles are kept closed and in the dark at 4 °C for at least 2-3 days. After this period, the seston will be deposited on the bottom of the sample bottle. The supernatant is removed by careful suction, using a plastic tube (that is not used for other purposes as Lugol's solution is adsorbed by the plastic). Hold the tube end just below the water surface, and reduce the volume to slightly less than 100 ml. Shake the bottle to resuspend the seston, transfer quantitatively to a 100 ml cylinder and make up to 100 ml with water. Cover the cylinder and repeat the procedure to reduce the volume to 10 ml.

## Storage:
Preserved material can be kept in closed brown glass containers and stored in the dark at 4 °C. If the samples are stored for a long period (months), neutralized formalin should be added to the Lugol preserved sample (1 ml of 40% formalin to 10 ml sample). For shorter storage periods some extra Lugol's solution should be added every two weeks to keep the colour dark brown.

## Analytical method(s):
*Method #1a.* Microscopical analysis, after concentration. The sedimentation method is simple in routine use and is adequate for diatoms, most small flagellates and some armoured dinoflagellates, provided the estuarine water is not too high in turbidity. The use of an inverted microscope is preferred. The concentrated sample is shaken, and a sub-sample of 3 ml is pipetted into a sedimentation chamber. After a minimum of 15 min deposition (but overnight deposition is better) it is sufficient to count between 20 and 100 of the units (cells, chains or colonies) of the more abundant taxa.

*Method #1b.* Microscopic analysis, non-concentrated sample. The counting is equal to Method #1a, the concentration step is not applied.

## Remarks:
Iodine, formalin and glutaraldehyde are toxic substances. Please refer to the safe handling procedures for chemicals.

**Method(s) references:**
Vollenweider (1969); Strickland & Parsons (1972); Sournia (1978); Throndsen (1978); Parsons *et al.* (1984); Tett (1987)

**A-7.42**

---

| **Variable:** | **Phytoplankton** |
| | |
| | **primary production** |
| | |
| **Unit:** | gC/m$^3$.d |
| | |
| **Compartment:** | water |

## Introduction:

The determination of primary production in waters gives a measure of the rate of growth of the phytoplankton community. Several techniques have been developed, which do not always agree well.

The performance of the measurements of primary production is based on two conceptually different approaches (Vollenweider, 1969):

a) discrete samples. Measurements that are carried out on isolated samples (either returned to the field or under controlled laboratory conditions). The latter method can be carried out under more standardized conditions, but it does not represent natural conditions (light, temperature) which should be simulated closely, however.

b) *in situ* measurements that are carried out directly in the environment, thus in non-isolated samples, hence no sampling is involved. The advantage is that the vertical gradient, which may be very important in estuaries because of the limiting effect of high turbidity, is taken into account. Also the natural conditions of the measurement are advantageous, but the relation to other (sampled and analyzed) variables measured in water samples may be difficult to interpret.

The most applied techniques use discrete samples and involve the uptake of a radio-tracer ($^{14}$C) by the growing plankton, or the production of oxygen.

As these involve batch techniques, and in many programmes the aim is to compare different variables as close as possible from one sample, these methods are preferred over the *in situ* methods.

The methods given below are usually carried out for relatively short incubation periods. The disadvantage of a short incubation is that diurnal variation in activity will not be included in the measurement. A 24 h incubation will not suffer from this problem. A 24 h incubation should preferably start soon after sunrise, so that all cells are in isotopic equilibrium. In the dark period the fixed $^{14}$C will subsequently be respired. Hence, a 24 h incubation will result in the net rate of photosynthesis, whereas a short incubation gives only a gross rate. Disadvantages of the 24 h method are that photo-adaptation may occur (especially at high light regimes), and that microbial processes continue which may lead to relatively large changes in biomass due to either growth (increase) or grazing (decrease in biomass).

For these reasons the methods presented below are only given for the short period incubations.

**Sampling:**
Sampling as for phytoplankton. As toxicants will hamper photosynthesis, non-metallic samplers should be used. Collection in acid-washed bottles to eliminate (trace metal) toxic effects; o-rings may contain toxicants as well.

**Sample volume:**
One litre will usually be sufficient.

**Sample treatment:**
Incubation should be carried out immediately after sampling in cleaned bottles. To protect the plankton from light-shock which may alter their response sub-surface samples should not be exposed to intense sunlight.

**Storage:**
Storage of the samples is not possible.

**Analytical method(s):**
*Method #1a.* $^{14}$C tracer technique. A known amount of radioactive carbonate ($^{14}CO_3^{2-}$) is added to a sea- or estuarine water sample of known carbonate content. After a specific time period (2-3 h) of photosynthesis by the phytoplankton under controlled conditions in an incubator (preferably with a range of light conditions) the plankton cells are filtered over a 0.45 μm membrane filter and the radioactivity from the carbon in the cells measured by (ß) scintillation counting. The uptake of radioactive carbonate, as a fraction of the total carbonate, will reflect the rate of net photosynthesis over the given period. Dark bottles results are used to be subtracted from the results obtained from bottles exposed in the light. They are thus corrected for dark uptake of $^{14}$C by bacteria and algae.
*Method #1b.* $^{14}$C tracer technique. As under Method #1a, but the incubation takes place *in situ,* by lowering the incubation bottles into the water column, preferably at the same depth.
*Method #2a.* Oxygen determination. The samples are incubated in dark and clear bottles for a specific time period (4 h) of photosynthesis by the phytoplankton under controlled conditions in an incubator. The evolution of the oxygen in the bottle is followed by oxygen determination (A-7.5). This can be by ion selective electrode, although the micro-Winkler titration is preferable because of its high sensitivity and the distinct ending of biological activity by the addition of chemicals (A-7.5). The dark bottle incubation will give the effect of respiration, the clear bottle the oxygen content due to the difference of respiratory consumption and primary production. From the increase in oxygen concentration the primary production is calculated.
*Method #2b.* Continuous oxygen determination. Instead of detection after a given period of time, the oxygen concentration is recorded continuously using an

oxygen electrode. From the concentration-time curve, the primary production is calculated.

**Remarks:**
$^{14}C$ is a (weak) radio-tracer which, in most countries will require a specially equipped laboratory space. Please check on safety regulations for working with radio-tracers.

**Method(s) references:**
Vollenweider (1969); Parsons *et al*. (1984)

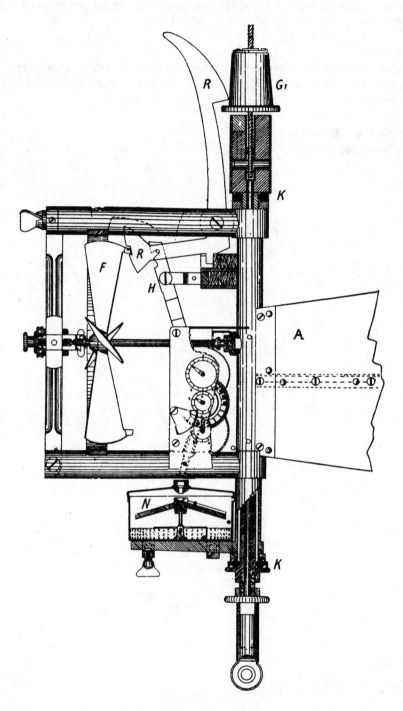

*The past:*
*The propeller current meter of Ekman (1905) (From: Krümmel, 1923)*

## A-7.43

**Variable:**        **Phytoplankton**

                     **biomass**

**Unit:**            **gC/m$^3$**

**Compartment:**     **water**

**Introduction:**

Phytoplankton is present in estuarine waters together with detrital and inorganic particles: seston (see S-6.2; A-7.27). The difficulty in analyzing phytoplankton is to separate living cells from non-living particles. This is important for the determination of phytoplankton biomass.

Biomass of phytoplankton is usually determined using conversion factors from chlorophyll data. Although quick and easy, the method has disadvantages. The ratio chlorophyll: biomass is not constant, ranging from 20 - 100 mgC/1 mg chlorophyll. Because of variation in cell size (and form) amongst species and physiological differences, correlations between chlorophyll and cell numbers are often weak. Additionally, along the salinity gradient chlorophyll may be present in dead cells and debris, even in fecal pellets.

Other methods involve estimating biomass from cell counts and cell size measurements, and the use of Coulter counter data. The latter method is not applicable in turbid estuaries, however. The latter method has the disadvantage that living phytoplankton can not be distinguished from dead cells and inorganic particles.

All methods are based on calculation rather than on direct measurement. It seems better to store the basic data (thus chlorophyll (A-7.29), species abundance (A-7.41) or Coulter Counter (A-7.28) data) in a database and perform the calculations in one place only, according to defined rules. Therefore the methods presented here are given an asterisk as a reminder that conversion factors are involved. Billington (1991) observed marked differences between the various methods, especially on a daily basis.

Example conversion factors from volume determinations are given in Annex V, while Jørgensen et al. (1991) have summarized several conversion relations. Cell sizes are necessary for a sufficient reliable conversion. Similar procedures can be applied to bacteria and zooplankton.

**Sampling:**

Sampling as for the species abundance.

**Sample volume:**

A volume of 1 litre is sufficient for either method.

**Sample treatment:**
See for the different methods under the respective analytical procedures (A-7.28, A-7.29, A-7.41). Preservation is not required.

**Storage:**
See respective analytical procedures (A-7.28, A-7.29, A-7.41).

**Analytical method(s):**
*Method #1\*.* Conversion from chlorophyll data.
*Method #2a\*.* Conversion from cell counts and cell measurements. The plankton cells are counted, their size is measured, and volume calculated.
*Method #2b\*.* Conversion from Coulter Counter data. From the total particle counts the total particulate volume is calculated by multiplying the counts for each channel by the respective particle volume, assuming equivalent spherical diameters.

**Remarks:**

**Method(s) references:**
Strickland & Parsons (1972); Sournia (1978); Parsons *et al.* (1984); Tett (1987)

**Variable:**      **Zooplankton**

               **species abundance**

**Unit:**          $n/m^3$

**Compartment:**   **water**


**Introduction:**
Members of the zooplankton vary from a few µm to several cm in size. Many
phyla have members that pass at least some part of their lives as plankton in the
water column. This heterogeneity of planktonic forms means that sampling
design and analytical procedures must be adapted to the group(s) of species
studied. In estuaries this problem is less acute than in open sea due to the
generally low diversity of the zooplankton. The methods described here are
applicable to the bulk of the zooplankton living in most estuaries.
From numbers and size of individuals (and of the various development stages)
and the (total) dry weight the biomass can be calculated.

**Sampling:**
The sampling device will vary according to the type of plankton to be sampled.
Water samplers or pumping systems are adequate for microplankton and
mesozooplankton, while towing nets are more specifically useful for collecting
large(r) species or when large numbers of organisms are required.
In many laboratories zooplankton sampling is carried out using towing nets
because of their relatively low price and easy handling. However, the use of
towing nets is seriously hampered in water with a high amount of suspended
particulate matter. In practice, it is almost impossible to sample quantitatively
turbid estuaries with towing nets with a mesh under 200 µm, due to clogging of
the net. The problem is avoided by using a net of a larger mesh-size, but small
zooplankton (*e.g.* nauplii, in case of copepod populations) are under-sampled.
While 200 µm mesh collected samples are suitable for rough comparisons
between estuaries, additional samples have to be collected by pumping system
or bucket (100 l over 55 µm mesh size net) for demographic and production
measurements. The micro-zooplankton will be sampled and analysed along with
the phytoplankton (A-7.41). Smaller samples may be required in this case.

**Sample volume:**
Using bucket, sampler or pumping system, 100 litres is preferred which is filtered
over a 55 µm gauze. With towing nets 5 $m^3$ is favoured (200 µm mesh).

**Sample treatment:**
One litre plastic flasks are convenient for transporting living material in order
to keep it cool. Net concentrated plankton should be copiously diluted with the
local estuarine water.
Zooplankton should be preserved in buffered formalin (4% formaldehyde) in
glass bottles, tightly closed and filled to the top. To prevent damage to the
organisms, the amount of plankton should not exceed 10% of the volume of
water.

**Storage:**
Preserved microplankton should be kept cool.
For zooplankton, preservation is improved if, after a few days of fixation, the
plankton is transferred to a mixture of 9 parts estuarine water and 1 part of the
following preservative: 50 ml propylene phenoxitol, 450 ml propylene glycol
and 500 ml of buffered 40% formaldehyde. Formalin is best buffered with borax
(excess or about 20 g/l); trimethylamine has also been used.

**Analytical method(s):**
*Method #1.* Microscopic analysis. Zooplankton samples containing much detrital and inorganic material are first treated with 1% aqueous solution of bengal
rose, staining the plankton pink, which facilitates counting.
Instead of, or prior to colouring, pre-separation of detritus and zooplankton can
be applied using a density gradient method (*e.g.* using Ludox™).
Sub-samples are taken in such a way that about 100 individuals are present in
the sub-sample. The examination can be performed using a Bogorov or a
Dollfus tray. All individuals of the sub-sample should be identified (including the
different development statges of copepods, see A-7.45) and counted.

**Remarks:**
Formalin is a toxic substance. Please refer to the safe handling procedures for
chemicals.

**Method(s) references:**
Tranter & Fraser (1968); Steedman (1976); Raymont (1983); Tett (1987)

**A-7.45**

---

**Variable:**        **Zooplankton**

**stage distribution (copepods only)**

**Unit:**           **n/m$^3$ (per stage)**

**Compartment:**    **water**

**Introduction:**
For a number of selected species, *e.g.* copepods, it is useful to have information
of the different development stages, the nauplii and copepodites. Usually 6
nauplii and 6 copepodite stages (N I - N VI and C I - C VI respectively) are
distinguished, of which the C VI is the adult stage.
In routine procedures the following development stages are identified and
counted:
- small nauplii (including N I - N II),
- large nauplii (including N III - N VI),
- all individual copepodite stages (C I - C IV),
- female and male (C V),
- and females and males (C VI).
Sampling, preservation and storage procedures are the same as those described
for the total species abundance (A-7.44).

**Sampling:**
Sampling according to the description under A-7.44 for total catch.

**Sample volume:**
Using bucket, sampler or pumping system, 100 litres is preferred, which is
filtered over a 55 μm gauze. With towing nets 5 m$^3$ is favoured (200 μm mesh).

**Sample treatment:**
As for the methods described under species distribution (A-7.44).

**Storage:**
As for the methods described under species distribution (A-7.44).

**Analytical method(s):**
*Method #1.* Microscopic analysis. Zooplankton samples containing much detrital and inorganic material are first treated with 1% aqueous solution of bengal
rose, staining the plankton pink, to facilitate counting. Instead of, or prior to
colouring, pre-separation of detritus and zooplankton can be applied using a
density gradient method (*e.g.* using Ludox™). Counting of sub-samples is as for
total abundance.

The examination can be performed using a Bogorov or a Dollfus tray. The entire sub-sample should be searched for the different stages of the selected species, which should be reported according to the N I -VI and C I - VI nomenclature. For the conversion to biomass size measurements of the different stages are needed.

**Remarks:**
Formalin is a toxic substance. Please refer to the safe handling procedures for chemicals.

**Method(s) references:**
Tranter & Fraser (1968); Steedman (1976); Raymont (1983); Tett (1987)

**A-7.46**

---

| **Variable:** | **Zooplankton** |
|---|---|
| | **stage weights (copepods only)** |
| **Unit:** | **µg (per stage)** |
| **Compartment:** | **water** |

**Introduction:**

Of the selected species, *e.g.* copepodes, the life stage weights are of interest amongst others for production calculations.

The analysis is carried out by weighing. It will depend on the number of stages present, how many subdivisions are found, and how many can reliably be weighed. It is anticipated that different results may be obtained when samples are stored deep-frozen or preserved in glutaraldehyde (or formalin, which may result in loss of some fat). Hence separate methods are defined here.

The results should be reported as dry-weights.

**Sampling:**

Sampling according to the description under A-7.44 for total catchment.

**Sample volume:**

Using bucket, sampler or pumping system, 100 litres is preferred which is filtered over a 55 µm gauze. With towing nets 5 $m^3$ is favoured (200 µm mesh).

**Sample treatment:**

The normal formalin preservation causes unpredictable loss of individual weights and makes this treatment unsuitable for biomass estimates.

Glutaraldehyde is a more effective fixative than formaldehyde because it cross-links proteins much more rapidly (Kimmerer & McKinnon, 1986). A 5% solution of glutaraldehyde in filtered seawater should be prepared in advance and maintained at or below room temperature. It should be used in a 1:1 dilution of the zooplankton sample.

**Storage:**

The preserved sample should be stored at or below room temperature. Dry-weight analysis should take place within a few days after fixation if possible. Zooplankton samples may be conveniently stored deep frozen if destined for chemical or biomass analysis.

**Analytical method(s):**

*Method #1a.* Glutaraldehyde (or formalin) preserved samples. Determination of dry-weight. The copepods are sorted and measured using a micrometer eyepiece fitted on a stereo-microscope. They are put into in groups of similar sized individuals, transferring them in a drop of water. In practice, size classes of about 100 µm wide are used corresponding to about 10 size classes for estuarine copepods (like *Eurytemora sp., Acartia sp.*) whose body lengths range from about 100 to 1000 µm. Each size class is subdivided into three batches (replicates) of at least 30 individuals (30 for the > 700 µm classes, 50 for the 400 to 700 µm classes and 100 for the < 300 µm classes).

The copepod batches are transferred to pre-weighed combustion boats by transporting them in a drop of water, after which the water is removed using a "fire thinned" glass pipette. This method also removes excess salt. The samples are dried for 24 hours at 60 °C. Until back-weighing, the samples are kept in a desiccator. Back-weighing using a microbalance with a precision of at least one µg gives the dry-weight per individual.

*Method #1b.* Deep frozen or fresh samples. After thawing, the procedure is identical to the method 1a.

**Remarks:**
Glutaraldehyde and formalin are toxic substances. Please refer to the safe handling procedures for chemicals.

**Method(s) references:**
Tranter & Fraser (1968); Steedman (1976); Raymont (1983); Tett (1987)

**A-7.47**

| | |
|---|---|
| **Variable:** | **Zooplankton** |
| | **biomass** |
| **Unit:** | **mg DW/m$^3$ (or mg C/m$^3$), (or mean stage length)** |
| **Compartment:** | **water** |

**Introduction:**

Zooplankton biomass estimates are crucial for secondary production assessment and should be carried out with great care and using standardized methods, allowing further comparisons between various studies and areas.

Essentially three methods are applied to estimate the zooplankton biomass: weighing (dry-weight), dry-weight estimation from length measurements and conversion from organic carbon content.

The following procedures mainly concern copepods which represent a dominant group in terms of zooplankton biomass and production.

Due to their small size, weighing copepods is a tedious job and it seems unrealistic to perform this as a routine analysis. It should be emphasized, however, that weighing gives the only true measured result (dry-weight) in contrast to the use of those calculated using conversion factors.

Being crustaceans with an exoskeleton, copepods have throughout their development, a rather standardized shape. This usually allows the establishment of log-linear regressions between the individual weight (W) and the length (L) with the function: $\log_{10} W = a \log_{10} L + b$ or $W = b * L^a$. When the parameters a and b have been determined for a population (calibration), the biomass is easily calculated from the length determination of the various development stages at each sampling date. For a mathematical treatment one is referred to Baskerville (1972), and to the summary in Annex VI.

As copepods are known to show size and shape variations according to seasons (a combination of thermic and trophic elements) it is safer to establish two regressions per year, one during spring (April) and one during autumn (October), which can be used for the summer-autumn and the winter-spring periods respectively. Copepods have then to be identified (development stages) and measured (cephalothorax length) at each (monthly) sampling interval and weighed only twice a year.

Biomass and derived production estimates are often expressed in carbon units. Despite the variations that have been found, the organic C/dry weight ratio is often assumed to be constant. Examples for a number of different zooplankton representatives are given, for example, in Parsons *et al.* (1984). Such conversion procedures should only be considered as a stopgap allowing comparisons to be made with other data expressed as carbon units and biomass should preferably expressed as mg dry weight/m$^3$ (*i.e.* the actually measured values).

Conversions from bio-volume to biomass is not recommended for estuarine zooplankton due to the usually large amount of detritus.

Biomass conversion equations have also bee collected by Jørgensen *et al* (1991). For more details see Annex V.

**Sampling:**
See A-7.44

**Sample volume:**
Usually, abundant material will be available for the determination of biomass. This means that a sample splitter has to be used to reduce the sample size to practical but still representative proportions for determination and analysis.

**Sample treatment:**
The normal formalin preservation causes unpredictable loss of individual weights and makes it unsuitable for biomass estimates. Glutaraldehyde is a more effective fixative than formaldehyde because it cross-links proteins much more rapidly (Kimmerer & McKinnon, 1986). The rapid reaction of glutaraldehyde causes proteins to become insoluble, which may prevent the loss of more labile organic substances.

A 5% solution of glutaraldehyde in filtered seawater should be prepared in advance and maintained at, or below, room temperature. The plankton sample, in a volume of seawater that is many times larger than the estimated settled volume of the plankton, is fixed by the addition of an equal volume of the dilute glutaraldehyde solution and gently agitated.

**Storage:**
The preserved sample should be stored at or below room temperature. If possible biomass analysis should take place within a few days of fixation.

**Analytical method(s):**
*Method #1.* Determination of dry-weight. After gentle rinsing with tap water the copepods are sorted under a micrometer eyepiece fitted on stereo-microscope. They are put into in groups of similar sized individuals, transferring them in a drop of water (see procedures under A-7.46). The copepod batches are transferred into pre-weighed combustion boats and dried for 24 hours at 60 °C or by freeze drying. Until weighing, the samples are kept in a desiccator. Back-weighing using a microbalance with a precision of at least one μg gives the dry-weight. For each size class the mean individual dry weight can be calculated, the summed dry-weight per $m^3$ is the total biomass.

*Method #2\*.* Dry-weight calculation from body lengths. As described under Method #1 the copepods are separated in size-classes and sub-samples. In each batch, the cephalothorax lengths are accurately measured under a microscope (x100 magnification) mounted with, for example, a drawing tube, ocular micrometer or digitizing tablet.

A regression has then to be found between the mean individual dry weights (cf. A-7.46) and the mean lengths (both log transformed) of the n size classes using the least squares method (see introduction). This log-linear regression between the individual weight (W) and the length (L) is described with the function: $\log_{10} W = a \log_{10} L + b$ or $W = b * L^a$. This equation can then be used to estimate the development stages mean individual weights from the length measurements proceeded on each sampling occasion.

It is advised for this Method that the actual measured data are preferred over the calculated biomass. This means that for each development stage the mean length and the mean weight (dry weight), with the numbers (n) and standard deviation (s.d.) of the data involved in the calculation of the mean have to be reported.

*Method #3\**. Conversion from organic carbon content. Biomass estimates are often expressed in carbon units. A useful technique in less turbid waters involves the direct analysis of the zooplankton of glass fibre filters (GF/C or GF/F) in a CHN analyzer (see A-7.30). A separation stepp of detritus and zooplankton may be necessary before the analysis. Despite the variations that have been found, the organic C:dry weight ratio is often assumed to be constant. The 0.5 ratio given by Heinle & Flemer (1975) is often used. Examples of conversion equations of different zooplankton representatives are given for example in Parsons *et al* (1984), different equations have also been collected by Jørgensen *et al* (1991). For more details see Annex V.

**Remarks:**
Glutaraldehyde and formalin are toxic substances. Both should be handled carefully, and high concentrations handled only under a fume hood. Please refer to the safe handling procedures for chemicals.

**Method(s) references:**
Heinle & Flemer (1975); Steedman (1976); Parsons *et al*. (1984); Kimmerer & McKinnon (1986); Jørgensen *et al*. (1991)

**A-7.48**

| | |
|---|---|
| **Variable:** | **Hyperbenthos** |
| | **species abundance** |
| **Unit:** | **n/m$^2$ or n/m$^3$** |
| **Compartment:** | **water** |

**Introduction:**
Members of the hyperbenthos vary from less than one mm to several cm in size.
The dominant hyperbenthic animals in estuaries are crustaceans (mainly mysids,
isopods and amphipods). Species belonging to many other taxa spend part of
their lives in the hyperbenthos (larval fish, crabs, carideans, etc.). The species
diversity in estuaries is generally low.

**Sampling:**
The lower 1 metre of the watercolumn should be sampled using devices with a
mesh size of 0.5 or 1 mm. Sledges are the prefered as sampling gear, although
other methods exist. See section S-6.7.

**Sample volume:**
In most estuaries one haul with a sledge will collect sufficient material for
analysis if conducted as follows: against the tide, 5 minutes at a ship speed of 1.5
knots; with the tide, 500 or 1000 metres at a ship speed of 4.5 knots.
Measurement of the volume of water filtered through the net and/or the distance
over which the sledge is in contact with the bottom, allows for conversion to
numbers per unit volume and/or area.

**Sample treatment:**
Hyperbenthos should be preserved in formalin (formaldehyde solution in
seawater, 10% final concentration).

**Storage:**
Formalin preserved samples can be stored until analysis. Deep freezing of fresh
material is not recommended because, after thawing, identification and
measurement of the animals may be difficult.

**Analytical method(s):**
*Method #1.* Visual inspection. Hyperbenthos samples containing a lot of detrital
and/or inorganic material are first treated with a 1% aqueous solution of bengal
rose, thereby staining the animals pink, which facilitates sorting. Samples are
rinsed over a sieve of a suitable mesh size (0.5 or 1 mm). Sub-samples are taken
in such a way that at least 500 individuals are present in the sub-sample. All

animals should be sorted out (easiest in a white tray), identified to species level and counted.

A binocular microscope (magnification x6 to x50) suffices for most species. For smaller species or difficult groups (*e.g. Gammarus spp.*) the use of a compound microscope may prove necessary. Zoea and megalopa stages of crabs as well as zoea and postlarval stages of caridean shrimp, as well as the different developmental stages of Mysidacea should preferably be distinguished (see section A-7.49).

**Remarks:**

Formalin is a toxic substance. Please refer to the procedures for the safe handling of chemicals.

**Method(s) references:**

Mees & Hamerlynck (1992)

**A-7.49**

**Variable:**        **Hyperbenthos**

                       **stage distribution (mysids only)**

**Unit:**               $n/m^2$ or $n/m^3$ (per stage)

**Compartment:**    water

**Introduction:**

For a number of selected species, *i.e.* Mysidacea, it is useful to have information on the length-frequency distribution of the different developmental stages present in the population. Usually 6 stages are distinguished: adult males, adult females, sub-adult males, sub-adult females, gravid females and juveniles. A seventh stage, post-gravid females, can sometimes be distinguished.

It is advisable to stage at least 200 individuals per species and per sample. At least 20 animals should be measured of each sex and stage.

**Sampling:**

As for the methods described under species abundance (A-7.48).

**Sample volume:**

As for the methods described under species abundance (A-7.48).

**Sample treatment:**

As for the methods described under species abundance (A-7.48).

**Storage:**

As for the methods described under species abundance (A-7.48).

**Analytical method(s):**

*Method #1.* Microscopic analysis. The preparation of the sample is similar to that for total abundance. At least 200 indivuduals are staged and sexed.

The following criteria can be applied to most estuarine species:

*adult males:* elongated 4th pleopods which reach beyond the posterior edge of the last abdominal segment, well developed and setose lobus masculinus between flagellae of antennal peduncle;

*adult females:* fully developed marsupium between thoracic legs;

*subadult males:* 4th pleopods stop short of reaching the end of the last abdominal segment; lobus masculinus present but small and not yet setose;

*subadult females:* small oostegites present;

*juveniles:* no secondary sexual characteristics;

*gravid females:* larvae present in marsupium.

Counting is as for total abundance. For the conversion to biomass, size measurements of the different stages are needed.

*Method #2.* Length-frequency distribution. These are obtained by measuring the length of at least 20 individuals per sex and per stage. The number of individuals within a stated length interval is then plotted against length. Lengths should preferably be measured as standard length, total length, or carapace length. Length classes are usually 1 mm if standard length or total length is measured; 0.5 mm if carapace length is measured.

**Remarks:**

Formalin is a toxic substance. Please refer to the procedures for the safe handling of chemicals.

**Method(s) references:**

Mauchline (1980)

**A-7.50**

---

**Variable:**     **Hyperbenthos**

**biomass**

**Unit:**     **mg AFDW/m$^2$ or mg AFDW/m$^3$**

**Compartment:**     **water**


**Introduction:**
When the total biomass of hyperbenthos samples has to be determined, weighing
of the entire sample is not advised. Conversion from length to weight is non-
destructive and is prefered as this technique keeps samples available for further
analyses. The use of ash-free dry-weight (AFDW) is the best choice
under estuarine conditions. Each species has to be treated separately. Accidental-
ly caught epibenthic species (*e.g.* adult shrimp or large demersal fish) should be
excluded from the analysis. Length-AFDW regressions should be determined for
a representative size distribution of all species with a more or less continuous
growth. Log-linear regressions can be determined between individual weight (W)
and length (L) with a function of the shape:
$\log_{10}W = a \log_{10}L + b$ or $W = b*L^a$. For a mathematical treatment of the length-
weight regressions we refer to Baskerville (1972), and to the summary in Annex
VI. The length-weight regressions may vary between and within estuaries
seasonally. For length measurements the use of standard length, total length or
carapace length is advised (conversion from still other measures are available
from literature).
For species growing in discrete stages (*e.g.* zoeae and megalopae of crabs)
average biomass values can be determined by weighing batches of animals.

**Sampling:**
As for the methods described under species abundance (A-7.48).

**Sample volume:**
A minimum of about 20 individuals should be analyzed in order to eliminate the
variability between organisms. For the Mysidacea, measurement of at least 20
individuals both per species and per stage is advised.

**Sample treatment:**
Formalin preservation causes unpredictable loss of individual weigths.
Optimally, fresh material should be used for the biomass determination, but if
analysis is performed after more than a few days, preservation becomes
necessary.

**Storage:**
(Preserved) samples should be stored at room temperature or under cooled conditions. Whenever possible, biomass analysis should take place within a few days after fixation.

**Analytical method(s):**
*Method #1.* Weighing (ash-free dry-weight, AFDW). The sample is dried at 60 °C to a constant weight (40-48 h), after which the sample is placed in a desiccator and allowed to cool. Individual animals can be dried for 2 hours at 110 °C. After weighing (using a microbalance with a precision of at least 1 μg) the sample is ashed at about 550 °C in a muffle furnace for 2-4 hours. After cooling to about 80 °C, the residue is stored in a desiccator. After weighing, the AFDW is calculated by substraction.
*Method #2*.* Ash-free dry-weight (AFDW) calculation from body lengths. Length should be accurately measured under a binocular microscope (magnification x12 or x25) mounted with a drawing mirror, ocular micrometer or digitizing tablet. For the calibration, a large number of individuals of all sizes should be used in order to correlate length and weight. A regression has then to be found between the individual AFDWs and the lengths (both log transformed) of the various size classes using the least squares method (see introduction). This equation can then be used to estimate the individual weights from the length measurements.

**Remarks:**
Formalin is a toxic substance. Please refer to the procedures for the safe handling of chemicals.

**Method(s) references:**
Crisp (1984)

**A-7.51**

| | |
|---|---|
| **Variable:** | **micro-phytobenthos** |
| | **species abundance** |
| **Unit:** | **n*10$^6$/m$^2$** |
| **Compartment:** | **sediment** |

**Introduction:**
In the sediment the inorganic sand and silt fractions will constitute the majority
of the particles. To obtain a quantitative insight in the distribution of the various
micro-phytobenthos species and their relative distributions, a separation
technique has to be applied.
Several methods have been used, which unfortunately give different results.
The most simple method is the use of lens tissue to harvest the mobile positive
phototactic micro-phytobenthos (Eaton & Moss, 1966). Other more elaborate
methods involve the use of density gradients, *e.g.* using a colloidal silica
suspension (*e.g.* Ludox$^{TM}$, De Jonge, 1979). The collection by means of different
sedimentation rates of sediment particles and microphytobenthos is not
recommended, because it appears that many of these organisms are attached to
the substrate.
For the identification of the species, microscopic investigation is indispensable.
This method is similar to that for phytoplankton.

**Sampling:**
Samples are collected by coring (about 2.4 cm ø). The top 5 mm is collected;
when deeper layers are to be examined, the layer 0.5-2 cm should be collected
additionally.

**Sample volume:**
To eliminate natural variability, about 20 individual samples are collected which
are pooled together (keeping the individual sediment layers).

**Sample treatment:**
Direct treatment of the sample is preferred. After separation of the micro-
phytobenthos from the sediment, the organisms have to be preserved in formalin
or in a Lugol's solution (see A-7.41).

**Storage:**
Lugol preserved samples can be kept in closed (brown) glass bottles and stored in
dark at 4 °C. If the samples are to be stored for several months, formalin should
be added (1 ml of 40% formalin to 10 ml sample).
As an alternative the preservation method of Van der Veer (1982) can be used.

**Analytical method(s):**
*Method #1.* Tissue technique. The sediment is spread in a petri dish, covered with lens tissue and placed in a climate room under illumination. Part of the mobile diatoms migrate into the tissue. After one day the tissue is rinsed with seawater and the organisms are collected and counted by microscopic inspection.
*Method #2.* Density gradient. Differences in the specific weight of micro-phytobenthos species, empty diatom frustules, organic detritus and inorganic sediment are used to separate the sediment in various fractions. A step-layered gradient is prepared of different density suspensions of colloidal silica polymers (*e.g.* Ludox$^{TM}$). Sub-samples of suspended sediment are added in the middle of the density gradient and centrifuged at medium speed. Afterwards the different fractions are harvested, and counted by microscopic inspection.

**Remarks:**

**Method(s) references:**
Eaton & Moss (1966); Cadée & Hegeman (1974); Van den Hoek *et al.* (1979); De Jonge (1979; 1992); Colijn & Dijkema (1981)

**A-7.52**

---

| **Variable:** | **micro-phytobenthos** |
| | **production** |
| **Unit:** | **gC/m$^2$.d** |
| **Compartment:** | **sediment** |

**Introduction:**
Primary production of the benthic microflora has been measured both in the
laboratory after sampling, and *in situ* using bell jars for treated field samples.
For primary production measurements only the top layer of the sediment is of
importance, despite the possible presence of a large part of the benthic microflora below this layer. The latter analytical method will give the total production
over the sediment column, which has been collected under natural conditions.
The method involving sediment sampling arbitrarily selects a sediment layer,
which is incubated *in situ* or in an incubator. Differences are therefore
anticipated.
As for phytoplankton, both the $^{14}$C incorporation method and the oxygen
production method can be applied.

**Sampling:**
For *in situ* determination using perspex cylinders (bell jars), no samples are
collected, but the core-tube is placed in the sediment on the studied location.
For incubation in bottles only the top 0.5 cm of the sediment is collected. The
collection of about 20 separate samples, which are pooled to get an averaged
sample, is recommended.

**Sample volume:**
For the bottle incubation methods 1 cm$^3$ is required.

**Storage:**
Storage of the samples is not possible, incubation should be carried out
immediately after sampling in cleaned bottles.

**Analytical method(s):**
*Method #1.* $^{14}$C tracer technique, *in situ,* in a bell jar. A (clear) perspex tube with
a lid (7-10 cm ø) is pushed into the sediment, either leaving some water or air
space. The core top part is completely filled with seawater. A known amount of
radioactive carbonate ($^{14}$CO$_3{}^{2-}$) is added to an estuarine water sample of known
carbonate content. After a specified time period (4 h) of photosynthesis by the
phytobenthos under controlled conditions in an incubator, the suspended cells are
filtered over a 0.45 µm membrane filter and the radioactivity from the carbon in

the cells measured by (ß) scintillation counting. The uptake of radioactive carbonate, as a fraction of the total carbonate, will reflect the rate of photosynthesis over the given period.

*Method #2a.* $^{14}C$ tracer technique, after sampling, incubation in the field. Sediment (sub)samples are placed in cleaned glass bottles, estuarine water is added. The analytical technique as under Method #1, the incubation takes place *in situ,* by placing the incubation bottles at the sediment surface, thus ensuring natural light and temperature conditions.

*Method #2b.* $^{14}C$ tracer technique, after sampling, incubation in an incubator. The analytical technique as under Method #2a, but the incubation takes place under controlled laboratory conditions by placing the bottles in an incubator.

*Method #3.* Oxygen determination, *in situ,* in a bell jar. The samples are incubated in either dark or clear perspex tubes inserted in the sediment for a specified time period (4 h) of photosynthesis by the phytobenthos under natural conditions. The change in concentration of the oxygen in the bottle is followed (A-7.5). This can be done by ion selective (micro)-electrode, although the micro-Winkler titration is preferable because of its high sensitivity and the distinct ending of biological activity by the addition of chemicals (A-7.5). The dark bottle incubation will give the effect of respiration, while the clear bottle will provide the oxygen content due to the difference of respiratory consumption and primary production. From the increase in oxygen concentration the primary production is calculated.

*Method #4a.* Oxygen production technique, after sampling, incubation in the field. Sediment (sub)samples are placed in cleaned glass bottles, and sea- or estuarine water is added. The analytical technique (oxygen determination) is as under Method #3. The incubation takes place *in situ,* by placing the incubation bottles at the sediment surface, thus ensuring natural light and temperature conditions.

*Method #4b.* Oxygen determination technique, after sampling, incubation in an incubator. The analytical technique as under Method #4a, but the incubation takes place under controlled laboratory conditions by placing the bottles in an incubator.

**Remarks:**
$^{14}C$ is a (weak) radio-tracer which in most countries will require a specially equipped laboratory space. Please check on safety regulations for working with radio-tracers.

**Method(s) references:**
Vollenweider (1969); Marshall *et al.* (1973); Cadée & Hegeman (1974; 1977); Colijn & De Jonge (1984); Round & Hickman (1984)

**A-7.53**

| | |
|---|---|
| **Variable:** | **micro-phytobenthos,** |
| | **biomass** |
| **Unit:** | **gC/m²** |
| **Compartment:** | **sediment** |

**Introduction:**
Chlorophyll *a* is not necessarily a good and conservative unit to describe the quantity of algal biomass. This is especially true for the estimation of microphytobenthos at the sediment surface, since here chlorophyll containing debris is collected (during slack tide) from the water column, which may thus cause an overestimation of the algal biomass. The determination of the microfloral biomass may be based on the organic carbon/chlorophyll *a* ratio (C/Chl-a). It appears that in time (seasonal effects) differences are found in this ratio. Fortunately, the ratio within an estuary for a given date appears to be constant (De Jonge, 1980). Differences in C/Chl-a ratios between various estuaries have not yet been recorded, but they may be expected, however.
A procedure that involves tuning of the method by using only two or three determinations of the ratio per sampling exercise along the estuary is presented here. The thus determined ratio can be used to estimate the biomass from the other stations (at that time and in that estuary) from the chlorophyll *a* determination only. This minimizes the number of organic carbon analyses, with its difficulties to separate microphytobenthos cells from other organic material in the sediment.

**Sampling:**
Samples are collected by coring (about 2.4 cm ø). The top 0.5 cm is collected; when deeper layers are to be examined; when deeper layers are to be examined, a second choice is the 0.5-2 cm layer.

**Sample volume:**
To eliminate natural variability, about 20 individual samples are collected and are pooled together (keeping the individual sediment layers); subsamples are taken from this homogenate for analysis.

**Sample treatment:**
The samples are lyophilized before analysis.

**Storage:**
The samples are to be stored deep-froozen at - 20 °C, or better.

**Sample volume:**
Depending on the expected chlorophyll content a volume of the extraction
liquid between 10 and 25 cm$^3$ is used for the determination of chlorophyll $a$.

**Analytical method(s):**
*Method #1.* Determination of the C/Chl-a ratio. After the sample water has been
removed, the sediment is froozen and lyophilized, and the chlorophyll $a$ content
determined spectrophotometricaly according to Lorenzen (1967). This and other
techniques have been described in A-7.29.
The organic carbon follows the procedure as described under POC (A-7.30).
The ratio (C/Chl-a) is calculated. Once the ratio is known, the biomass for all
other stations can be estimated from the Chl-a determination alone.
For the description of the analytical methods one is referred to A-7.51.

**Remarks:**
Remember that not only vital micro-phytobenthos cells are represented by the
chlorophyll $a$ measurement (De Jonge, 1992).

**Method(s) references:**
Lorenzen (1967); De Jonge (1980; 1992); Round & Hickman (1984)

**A-7.54**

| | |
|---|---|
| **Variable:** | **macrophytobenthos** |
| | **species abundance** |
| **Unit:** | **g C/m$^2$ (DW) (per species)** |
| **Compartment:** | **sediment** |

**Introduction:**
For many programmes the species abundance is important for general classification. The macroscopic morphology which predominantly separates the different
genera determines the classification. The qualitative and quantitative aspects of
the species composition will be carried out in the laboratory rather than in the
field.
As numbers per m$^2$ give no relevant information, the species composition
should be given in terms of either biomass per species or as percentage of the
total biomass. The latter may be calculated from the former and therefore
biomass (dry-weight) per species is preferred.

**Sampling:**
The macrophytobenthos is quantitatively collected from the randomly selected
quadrats as defined under S-6.9. The material is kept in plastic bags.

**Sample volume:**
All material within each selected quadrat should be sampled for the determination of the species distribution.

**Sample treatment:**
The macrophytes are cleaned of adhering particulate matter and are kept moist;
excess water should be avoided.

**Storage:**
The plastic bags should be stored in a cool place (4 °C), in the dark. The
classification should be performed within 24 hours.

**Analytical method(s):**
*Method #1.* The macrophytes are to be identified at least to the level of
macroscopic morphological characteristics. The numbers of each representative
species and their wet-weight should be recorded, their biomass (dry-weight)
determined (see A-7.55). This enables the calculation of their relative
distribution.

**Remarks:**

**Method(s) references:**
Wolff (1983; 1987b); Eleftheriou & Holme (1984); Round & Hickman (1984)

**Variable:**        **macrophytobenthos**

                   **total biomass**

**Unit:**           **g/m$^2$ (DW)**

**Compartment:**    **sediment**

**Introduction:**
For the determination of the biomass of macrophytobenthos, the collected
material is dried to determine the dry-weight. Large quantities are sometimes
collected from the selected quadrats of substrate and sub-sampling may become
necessary. The sub-sampling procedure should then be checked carefully to
ensure collection of representative amounts are collected. This can be achieved
by subsequent determination of total wet-weight of the sample (preferably per
selected class), and the wet-weight of the sub-sample. The dry-weight of the
latter is then determined. The total biomass should equal the sum of all different species biomass as determined for the species abundance.
The wet-weight might be measured which can be converted to dry-weight
biomass by calculation. However, as this requires non-standard conversion
factors the method is not recommended and hence this method has an asterisk.
It should be mentioned that also remote sensing techniques have been applied
for the estimation of the biomass.

**Sampling:**
Because of the high variability with time of the standing stock of macrophytes
(growth and erosion) Wolff (1987a,b) stresses the need for frequent sampling.
The macrophytobenthos is quantitatively collected from the randomly selected
quadrats, as defined under S-6.9. The material is sorted for the major macroscopic morphological characteristics (genera).

**Sample volume:**
The total material per sampling square is wet-weighed; sub-samples are prepared when necessary, to get about 100 g wet-weight of material for dry-weight
analysis.

**Sample treatment:**
To avoid weighing errors from non-macrophyte material, the samples are
carefully washed to remove adhering particles and organisms that may have
been collected as well. After sorting into separate classes and sub-sampling
when necessary, the macrophytes are kept in plastic bags.

**Storage:**
To prevent deterioration of the material, the samples should be stored moist (not wet) in a cool place (4 °C) in the dark. The analysis should take place within 24 h.

**Analytical method(s):**
*Method #1.* About 100 g wet material of each macrophyte class, and of each substrate square, is dried at 80 °C for 24 h. The dry-weight is averaged from the various squares, and reported as dry-weight per class of macrophytes and calculated per m$^2$.
*Method #2\*.* If the macrophytes are to be used for further analysis, the wet-weight is determined per class and per sampling square, after blotting with tissue paper. The results are expressed as wet-weight per class of macrophytes, and calculated per m$^2$. Conversion to dryweight by assuming a constant factor (per species).

**Remarks:**

**Method(s) references:**
Wolff (1983); Round & Hickman (1984)

**A-7.56**

| | |
|---|---|
| **Variable:** | **macrophytobenthos** |
| | **trace metal content** |
| **Unit:** | **mg/kg** |
| **Compartment:** | **sediment** |

**Introduction:**

Aquatic macrophytes accumulate significant amounts of trace metals from ambient waters and the concentrations of elements present in tissues of these plants can proveide information on the extent of contamination of estuarine and coastal waters (Phillips, 1994). Especially species of brown and green algae, but also seagrasses have thus been used. In contrast to the many trace metal studies, macrophytes have only rarely been used to quantify the abundance of trace organic contaminants, which is probably due to the low lipid content of these organisms.

Usually only one or a few species are selected and used for trace metal analyses, *e.g. Fucus* spp., *Enteromorpha* spp.).

For pollution studies it is very important that the surfaces of the plants are washed carefully, as both organic (*e.g.* epiphytes) and inorganic (adhered particulate matter) contamination will otherwise occur.

For trace metal studies separate samples are to be collected in addition to those for biological analysis, taking care not to contaminate the tissue.

Analysis for trace metals can be sub-divided into two steps: the destruction step (digestion) and the instrumental analysis.

Wet digestion techniques are generally employed for most elements, using nitric acid or mixtures of acids, like nitric and perchloric or nitric and sulphuric acids (Say *et al.*, 1986). Although most methods seem to work well, small differences may occur. The method should be validated for the species to be analysed.

Instrumental analytical methods for trace elements compare reasonably well. Although large discrepancies may exist between laboratories (which is a cause for concern), no systematic differences have been obeserved between the techniques proper. Commonly applied are various atomic absorption methods (flame AAS, electrothermal AAS, cold vapour AAS (for Hg)); when multi-elements are to be analysed, inductively coupled plasma mass spectrometry (ICP-MS) or neutron activation analysis (NAA) may become advantageous. These techniques involve costly instruments, however. Differences in sensitivities occur, and the method should be validated for the elements (and species) tested.

**Sampling:**

Selected species are collected in areas adjacent to the randomly selected quadrates selected (see S-6.9; A-7.54). Care should be taken not to contaminate the sample. No metal tools should be used during collection or processing. Correlations have been found beteen different tidal horizons and trace metal content, and this should be taken into consideration when developing the sampling strategy.

The macrophytes may be pre-cleaned of adhering epiphytes and particulate matter at the sampling site, using the local seawater. They are kept moist in a plastic bag and transported under cool (4 °C) conditions to the laboratory.

**Sample volume:**
To reduce natural variability, a minimum of 25 inividuals (from each horizon) need to be collected, which may be pooled into one combined sample, however.

**Sample treatment:**
Even when samples were pre-cleaned in the field, the macrophytes need to be cleaned of adhering particulate matter in the laboratory. Unfortunately no standard method has been developed yet, but rinsing with clean filtered seawater or distilled water are most used methods, sometimes with the assistance of an ultrasonication.

After cleaning of the surface the samples are homogenised. The homogeniser should not contain metal parts that are in contact with the sample (or the metal blades are to be of a non-interfering metal, like titanium).

**Storage:**
When analysis (destruction) is carried out within days, the samples are to be stored cool (4 °C), otherwise the may be stored deepfrozen until chemical analysis. Storage containers should be of plastic (polythene, polypropylene).

**Analytical methods:**
*Method #1a.* Digestion by $HNO_3$, followed by AAS detection;
*Method #1b.* Digestion by $HNO_3/HClO_4$, followed by AAS detection;
*Method #1c.* Digestion by $HNO_3/H_2SO_4$ followed by AAS detection;
*Method #1d.* Digestion by other acid mixture, followed by AAS detection;
*Method #2a-d.* Similar digestions, followed by other instrumental detection;
It is impossible to cover all possible combinations for the digestion/analysis of macrophytes. After digestion a clear solution should result. As the AAS method is most commonly available only this method is split up into sub-methods. Both digestion and analysis should be validated; (certified) reference material should be incorporated to test for analytical procedures (Cantillo, 1992).

**Method(s) references:**
Say *et al.* (1986); Freitas *et al.* (1993); Phillips (1994)

**A-7.57**

| | |
|---|---|
| **Variable:** | **meiobenthos** |
| | **species abundance** |
| **Unit:** | **n/m$^2$** |
| **Compartment:** | **sediment** |

**Introduction:**
For the determination of the species distribution of a core sample, it is essential that a proper separation between the (inorganic) sediment and the meiofaunal organisms is quantitatively achieved. Furthermore, the number of species to be analyzed should be such that sufficient organisms are present to get a reliable estimation of their relative contributions, and that the amount of work is kept within reasonable limits. The sampling method (core diameter, core length) tries to cope with the latter aspect.

The methods of retrieval of the organisms from the sediment has extensively been treated by Pfannkuche & Thiel (1988), comparisons have been made by Uhlig *et al.* (1973).

In general the sediment is gently sieved under water to separate the animals from the abiotic particulate fraction. The following mesh sizes are proposed: 1000 μm (upper limit of meiofauna); 500 μm (mainly consisting of juvenile macrofauna, temporary meiofauna); 250 μm; 125 μm; 63 μm and 45 μm (lower limit for the retention of meiofauna). The larger sieve meshes are in principle used to separate several size fractions. The lower limit is important as it determines the minimum size that is retained, and thus included in the meiofauna analyses. The 45 μm mesh size is preferred as the lower limit, sometimes the 63 μm lower limit is used for practical reasons.

Several methods for the quantitative extraction and concentration of meiofauna specimens have been described, none of which seems to be the ultimate technique. These rely on a) differences in the sinking rates of sediment particles, detritus and meiofauna (decantation, elutriation, flotation, filtration over a density gradient) and b) response (of living organisms) to environmental gradients (oxygen depletion, temperature, phototactic reaction, electro-shocking). Also density gradients, *e.g.* using Ludox$^{TM}$ have been proposed (de Jonge & Bouwman, 1977).

After quantitative retrieval of the organisms the quantity may be too large for identification of all individuals. Splitting techniques are useful in these cases.

**Sampling:**
The various sampling techniques of sediment for the collection of meiofauna, either in sub-tidal or inter-tidal areas, in sandy sediments or in mud, is described

in section S-6.10. Standard sampling depth is 5 cm for muddy sediments, and 15 cm for sandy sediments.

**Sample volume:**
The sample volume is highly related to the type of sediment, and hence the number of meiofauna species to be expected. This factor determines different sampling strategies for the various substrates (S-6.10). Splitting of the sample (core) in 2 or 4 equal parts may be a possibility to reduce the sample size (check the equal splitting by weighing and counting).
Sub-sampling by splitting of the sediment extracted organisms may performed using various types of devices (Pfannkuche & Thiel, 1988).

**Sample treatment:**
(Sediment) samples should be stored after addition of 4% warm formalin (60 °C) in seawater solution. The elevated temperature stops nematodes rolling up, which facilitates their identification.
After extraction, identification is facilitated by staining of the entire sample with rose bengal (1 %, 48 h) or fluorescent dyes. The fauna is extracted, depending on sediment type, according to methods described in detail in Pfannkuche & Thiel (1988).

**Storage:**
Preserved samples can be stored until analysis.

**Analytical method(s):**
*Method #1.* Microscopic analysis. The meiofauna may be identified to major taxa and enumerated under a stereomicroscope. For more detailed taxonomic analysis organisms should be sorted into major taxa and treated according to the relevant sections in Higgins & Thiel (1988).

**Remarks:**
Formalin is a toxic substance. Please refer to the safe handling procedures for chemicals.

**Method(s) references:**
Uhlig *et al.* (1973); De Jonge & Bouwman (1977); Bouwman (1987); Higgins & Thiel (1988); Pfannkuche & Thiel (1988)

# A-7.58

**Variable:**    **meiobenthos**

**biomass**

**Unit:**    **g/m$^2$**

**Compartment:**    **sediment**

**Introduction:**
Two methods that differ substantially are used for the determination of the
biomass: the gravimetric and the volumetric methods.
When large numbers of meiofaunal organisms are available their biomass can be
determined directly from the dry-weights.
The second method involves the "measurement" of the total volume of the
various taxa. This measurement is through visual inspection of the specimens and
subsequent estimation of their body volume by calculation using simple formulae
for regular shaped ones (*e.g.* nematodes), length to volume conversions using
preset conversion factors, and/or the use of clay models for irregularly shaped
species or the use of slides at known equidistance for soft irregular forms. The
conversion to biomass is thus derived from a calculation involving an arbitrary
specific gravity (usually of 1.13) (Feller & Warwick, 1988).
While direct gravimetric measurement gives a straightforward answer, the other
methods use conversion factors that may change with time or investigator. An
example is the volume to dry-weight conversion factor of 1.13 (specific gravity),
and a dry-weight to wet-weight ratio of 0.25. They have therefore been marked
with an asterisk. For the JEEP92 project the methods as given in Feller &
Warwick (1988) have been chosen. Details of the conversion factors from this
reference are given in Annex V.
It is important to have the actual measured values instead of data that are derived
from unsure conversion to be stored in a database to allow application of any
desired conversion later. Therefore dry-weight, ash-free dry-weight or blotted
wet-weight are accepted. The first has, however, our preference.

**Sampling:**
The various sampling techniques of sediment for the collection of meiofauna,
either in sub-tidal or inter-tidal areas, in sandy sediments or in mud, is described
in section S-6.10. Standard sampling depth is 5 cm for muddy sediments, 15 cm
for sandy sediments.

**Sample volume:**
The sample volume is highly related to the type of sediment, and hence the
number of meiofauna species to be expected. This factor determines different
sampling strategies for the various substrates (S-6.10). Splitting of the sample

(core) in 2 or 4 equal parts may be a possibility to reduce the sample size. Check the splitting of the sediment by weighing or preferably counting of organisms.

**Sample treatment:**
(Sediment) samples should be stored deep frozen (-20 °C) or after addition of 4% (warm, 60 °C) formalin in seawater solution. Substantial differences in biomass have been found between preserved and non-preserved samples (Widbom, 1984) The method of retrieval of the organisms from the sediment has been extensively treated by Pfannkuche & Thiel (1988). Several methods for the quantitative extraction and concentration of meiofauna specimens has been described, none of which seems to be the ultimate technique. These rely on a) differences in the sinking rates of sediment particles, detritus and meiofauna (decantation, elutriation, flotation) and b) response (of living organisms) to environmental gradients (oxygen depletion, temperature, phototactic reaction).

**Storage:**
Well preserved samples, either under formalin or deep frozen can be stored for a reasonable time until analysis.

**Analytical method(s):**
*Method #1a.* Gravimetric determination of dry-weight, formalin preserved. When large numbers and/or large specimens are available, dry-weights can be determined directly on appropriately sensitive balances (sensitivity $\pm 0.1$ µg). Prior to weighing, the organisms are rinsed by washing over a GF/C glass fibre filter and dried in an oven at 60 °C until constant weight (24 h). The should be stored in a desiccator until weighing. The biomass is given as g dry-weight per $m^2$.
*Method #1b.* Gravimetric determination of dry-weight, preserved deep frozen. The method is identical to Method #1a, except for the preservation.
*Method #2a.* Gravimetric determination of ash-free dry-weight, formalin preserved. As Method #1a, but after drying, the sample is ashed in an oven at 500 °C for 6 h, and expressed as g ash-free dry-weight per $m^2$.
*Method #2b.* Gravimetric determination of dry-weight, preserved deep frozen. As Method #1b, but ashed in an oven and expressed as g ash-free dry-weight per $m^2$.
*Method #3a.* Gravimetric determination of blotted wet-weight, formalin preserved. As Method #1a, but blotted dry using tissue paper, and expressed as g g blotted wet-weight per $m^2$.
*Method #3b.* Gravimetric determination of blotted wet-weight, preserved deep frozen. As Method #1b, but blotted dry using tissue paper, and expressed as blotted wet-weight per $m^2$.
*Method #4\*.* Volume based conversions. Depending on the shape of the species the volume is calculated from simple equations involving length and diameter (*e.g.* worms) or using empirical conversion factors that take the irregular shape into account (see Annex V).

**Remarks:**
Formalin is a toxic substance. Please refer to the safe handling procedures for chemicals.

**Method(s) references:**
Widbom (1984); Feller & Warwick (1988); Greiser & Faubel (1988); Pfannkuche & Thiel (1988)

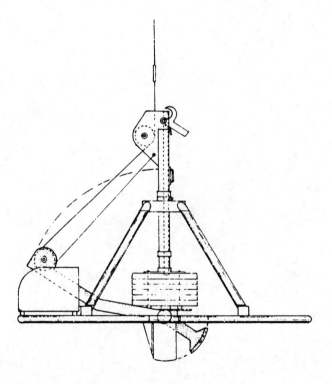

*Figure 8.*
*The Reineck boxcorer (From: Reineck, 1963)*

**Variable:**  **macrozoobenthos**

**species abundance**

**Unit:**  **n/m$^2$**

**Compartment:**  **sediment**

**Introduction:**
Macrofauna is distinguished from the meiobenthos by arbitrary selection of mesh size of the sieve of 1 mm; sometimes 0.5 mm is preferred. The sediment sample is washed over a sieve with a gentle water stream to remove sediment and smaller organisms. Specialized techniques involve a freezing and Calgon technique (sodium hexametaphosphate for breaking up stiff clays), an elutriation technique (floatation in a liquid with higher density than water), or the use of the swimming behaviour of organisms (Hartley *et al.*, 1987).
For identification of the individual species, one is referred to the overview of the literature concerning this subject (Wolff, 1987b).

**Sampling:**
The use of corers is preferred to grab samplers because of the better definition of the surface area and sediment sampling depth, thus a more consistent sample size. Their use will be dependent on the sampling site (sub- or intertidal), and on the availability of a suitable ship (*e.g.* when using box-corers).
Sampling depth should be 25 cm, replicate samples should be collected (S-6.11). It has been found that benthic zonation may influence macrofauna related variables. Therefore three sites: upper, mid and lower tidal level are ideally to be sampled. It will be clear that the number of replicates should be applied to each of these spots.

**Sample volume:**
In most estuaries one core of, for example, 200 cm$^2$ surface area, will collect sufficient material for analysis. In sediments with high organism densities smaller cores will be sufficient. Due to heterogeneity in the distribution of the organisms in the sediment, a series of individual samples (= replicates) should be collected and analyzed separately. This will yield statistical information, *e.g.* about the patchiness of the macrofaunal distribution.

**Sample treatment:**
After extraction over a 1 mm mesh (or 0.5 mm) the organisms should be preserved with buffered formalin (6% formaldehyde in seawater; but if the samples contain a lot of water, 8-10% is better). Bulk staining has been proposed

to facilitate sorting. A stain, like bengal rose or eosin can be added to the formalin solution.

**Storage:**
Formalin preserved samples can be stored until analysis. Deep freezing of fresh material is not recommended because after thawing identification of the organisms may be difficult. Fresh samples (up to a few days, stored cool at 4 °C) are sometimes easier easy to sort (moving animals) and to identify. Storage of formalin preserved samples deep froozen has been recommended as a good alternative (Salonen & Sarvale, 1985).

**Analytical method(s):**
*Method #1.* Visual inspection. The samples are placed in a white tray and checked for species which are identified. The small sized animals should be identified using a binocular microscope and possibly a compound microscope.

**Remarks:**
Formalin is a toxic substance. Please refer to the safe handling procedures for chemicals.

**Method(s) references:**
Barnet (1979); Coleman (1980); Eleftheriou & Holme (1984); Warwick (1984); Hartley *et al.* (1987)

**A-7.60**

| | |
|---|---|
| **Variable:** | **macro-zoobenthos** |
| | **age distribution** |
| **Unit:** | **n/class** |
| **Compartment:** | **sediment** |

**Introduction:**
For some important species (especially molluscs) the age composition is aquired. Distinction of age can be based on either growth marks on some hard part of the organism (shells of bivalves, jaws of polychaetes) or on size-frequency distributions. The latter method becomes problematic with the older (slower growing) individuals because of overlap between the fast growing of one year and the slow growing of the next year.

**Sampling:**
The same as for the total species distribution. As part of the sorting procedure the species for which the age-classes are to be determined may be kept aside for further study (sub-sampling).

**Sample volume:**
The same as for total species distribution.

**Sample treatment:**
After extraction the organisms should be preserved with buffered formalin (6% formaldehyde in seawater).

**Storage:**
Formalin preserved samples can be stored until analysis.

**Analytical method(s):**
*Method #1.* Growth marks. This method is suitable provided that it can be proved that growth marks are laid down annually in *e.g.* the shell of a mollusc or in the jaws of a polychaete, and that they can be counted. One must be able to distin-guish between growth checks produced by the slowing down of somatic growth during winter and those induced by diversion of energy into the production of gonads (seasonal effects) or caused by (accidental) adverse physical conditions *e.g.* prolonged low oxygen tension.
The year marks are counted and the organisms are place in the respective age classes.
*Method #2.* Frequency distribution. These are obtained by measuring the length (or weight) of a large number of individuals, preferably the entire sample and

plotting the number of individuals within a stated size interval against size. A series of peaks can usually be distinguished.

Graphical methods may be used to separate peaks in a size/frequency distribution (Harding, 1949; Cassie, 1954). While the first two peaks generally relate to discrete age classes there is often considerable doubt as to the number of age classes in any additional peaks. If one can estimate the number of age classes that there should be in a sample there are a number of methods (*e.g.* McDonald & Pitcher, 1979) by which complex size histograms can be composed to give age classes.

**Remarks:**

Formalin is a toxic substance. Please refer to the safe handling procedures for chemicals.

**Method(s) references:**

Cassie (1954); Bhattacharya (1967); MacDonald & Pitcher (1979); Crisp (1984); Warwick (1984)

**A-7.61**

| | |
|---|---|
| **Variable:** | **macrozoobenthos** |
| | **biomass** |
| **Unit:** | **g/m$^2$ (AFDW)** |
| **Compartment:** | **sediment** |

**Introduction:**
Different methods are used to give a representation of the biomass estimation.
They can be expressed on the basis of wet-weight, dry-weight, ash-free dry-
weight or in terms of total N-content or caloric value. It will be obvious that
differences between the techniques occur. Conversion factors have been
evaluated for several relations between these measures which indicates that
conversions are not universally applicable either in space or in time.
When the organisms contain a substantial amount of sediment in the gut (*e.g.*
polychaetes), defecation in clean seawater for 24 h is essential to overcome errors
in the biomass determination by weighing technique.
Another approach is a conversion from length (or volume) to weight. This
involves either an arbitrary factor, or can be based on actual measurements from
the estuary and relevant period of the year (calibration). Since macrozoobenthos
is usually of sufficient size to determine the actual weight, weighing is preferred
to the conversion method. Although biomass determination can be performed on
(homogenised) sub-samples, the non-destructive method is preferred, as it keeps
the samples available for further analyses, *e.g.* histo-pathology, chemical analysis of trace elements.
The use of ash-free dry-weight is the best choice under estuarine conditions.
This eliminates errors caused by fluctuations in the internal salt content which
varies according to estuarine conditions. As the other methods are also widely
applied, they are given as well.
For animals with hard parts, such as mollusca, a length-weight conversion can
be calculated. A log-linear regression is determined between the individual
weight (W) and the length (L) with the function of the shape:
$\log_{10}W = a \log_{10}L + b$ or $W = b * L^a$. For a mathematical treatment of the length-
weight relations, one is referred to Baskerville (1972), and to the summary in
Annex VI. When the parameters a and b have been determined for a population
(calibration), the biomass is easily calculated from the length determination. This
weight-length relation may vary with the month of the year, and should therefore
be checked. The regression will differ between estuaries and since the condition
of the organisms also changes between seasons it is advisable to calibrate the
method accordingly for each estuary studied.

**Sampling:**
As for the species distribution (A-7.59).

**Sample volume:**
A total of about 0.5 - 3 g of (wet) tissue, depending on the balance used, is sufficient for an (ash-free) dry-weight determination. To eliminate the variability between organisms a minimum of 20 individuals should be analyzed, either by homogenisation (and subsequent sub-sampling), or by separate analysis and calculation.

**Sample treatment:**
For the biomass determination the samples should preferably not be treated with formalin or any other fixing agent, as these may interfere with the analysis. Optimally fresh material should be used. As an alternative the material should be stored deep frozen (- 20 °C) until analysis, preferably after sub-sampling for species and age-class. Freezing individual animals tends to cause weight loss, however. Additionally material may be lost when tissues rupture on thawing. Biomass is usually determined on the soft tissue. For the determination of the biomass of invertebrate taxe with an exoskeleton (*e.g.* molluscs, echinoderms) this non organic matter should be removed before the DW is determined. This can be by simply preparing the tissue out, but also by decalcifying using 10% HCl. The acid has to be renewed until complete dissolution of the calcareous parts (shell, ophiurid vertebrae, echinid test, etc.) has been achieved. The tissue should be rinsed with water before drying. One should take care not to loose organic matter in the process of renewal.

**Storage:**
The samples should be used fresh or stored in plastic bags or containers, deep frozen (- 20 °C).
Bulk freezing of sediment samples containing the animals should be avoided at all cost.

**Analytical method(s):**
*Method #1.* Weighing (ash-free dry-weight). The sample is dried at ca 60 °C to constant weight (40-48 h), after which the sample is placed in a desiccator and allowed to cool. The sample is weighed, incinerated at about 550 °C in a muffle furnace for 2-4 h. After cooling to 80 - 90 °C, the residue is stored in a desiccator. After weighing the ash-free dry-weight is calculated by subtraction.
*Method #2.* Weighing (dry-weight). The sample is dried at ca 60 °C to constant weight (40-48 h), after which the sample is placed in a desiccator and allowed to cool. The sample is weighed to determine the dry-weight.
*Method #3.* Weighing (blotted wet-weight). The sample is blotted dry on paper tissue and weighed to determine the wet-weight. From this wet-weight the biomass can be estimated by the use of conversions. This method is relatively fast and allows the use of the organisms for other analyses. The method should be

calibrated using the formula: $\log_{10} W = a \log_{10}(\text{wet-weight}) + b$ , where W is the dry-weight.

*Method #4\**. Dry-weight calculation from body lengths. For organisms that are suitable for the method, *i.e.* containing hard parts, the lengths are accurately measured. For the calibration a large number of individual organisms of all sizes should be used to correlate length and weight. A regression has then to be found between the individual ash-free dry weights (cf. Method #1) and the lengths (both log transformed) of the various size classes using the least squares method (see introduction). This equation can then be used to estimate the individual dry-weights from the length measurements.

*Method #5\**. Dry-weight calculation from body volume. This method is important for irregular shaped organisms, where the length is not suitable for conversion purposes. As for the length - dry-weight conversion, there is no constant relationship between volume and tissue dry-weight, not even for a given species. The method should therefore be calibrated using the formula: $\log_{10} W = a \log_{10}(\text{volume}) + b$ or $W = b * (\text{volume})^a$ , where W is the dry-weight.

**Remarks:**

Formalin is a toxic substance. Please refer to the safe handling procedures for chemicals.

**Method(s) references:**

Crisp (1984); Warwick (1984)

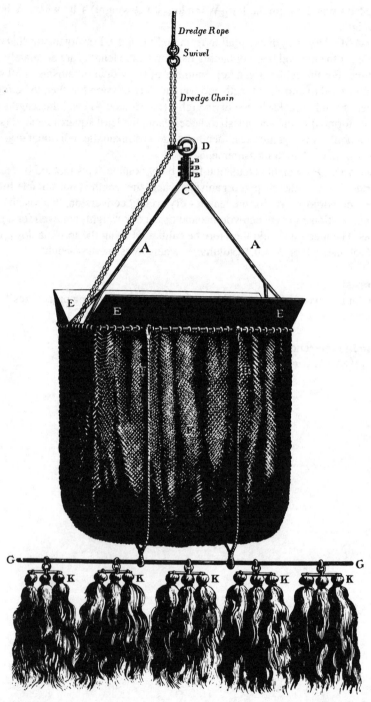

*The past:*
*The benthic dredge used by the Challenger expedition (1873-76) (From: Challenger*
*Report, 1885)*

**A-7.62**

| | |
|---|---|
| **Variable:** | **macrozoobenthos** |
| | **trace metals** |
| **Unit:** | **mg/kg** |
| **Compartment:** | **sediment** |

**Introduction:**

Some species from the macrozoobenthos are often used as bioindicators. They accumulate pollutants, including trace metals, in their tissue, reaching an equilibrium with the ambient concentration. The tissue/water ratio is known as the bio-concentration factor (BCF). A good impression of the ambient concentrations, collected by the organisms over a relatively large period (integrative sampling) may be gained by chemically analyzing the tissue. The international 'Mussel Watch' type studies are examples of this type of approach (Cantillo, 1991; O'Connor *et al.*, 1994; De Kock & Kramer, 1994).

Bivalves lend themselves very well to the biomonitoring approach with mussels (*Mytilus* spp., *Perna* spp.), oysters (*Ostrea* spp., *Crassostrea* spp.) and clams often being used. Other benthic organisms have also been used in monitoring studies *e.g.* worms (*Arenicola marina*, *Nereis* spp.), crustacea (shrimps, barnacles, crabs), echinoderms, etc. It is important to identify the individuals to the species level, as large differences may occur, even between apparently similar organisms.

To avoid natural variation becoming dominant in the analysis, a relatively large number of individuals need to be collected, then homogenised and pooled into one composite sample. In order not to disturb the analysis, organisms which contain much sediment in their stomachs (worms) need to depurate for 24 h in clean seawater before being homogenised. For bivalves this is still a point of discussion, but depuration is not usually applied.

Contamination control is of the utmost importance: no metal tools should be used during sampling or sample preparation (unless the metal is not critical in the analysis: we use titanium blades for dissection and homogenization of the tissue). As not all organisms can be digested using the same method, no agreement has been reached on a common method of digestion. Acids ($HNO_3$, $HClO_4$, $H_2SO_4$) alone or in combination, sometimes with peroxide ($H_2O_2$), are used. When digestion techniques prove similarly efficient, the analytical technique will determine the final choice (matrix effects, detection limits, blanks). Any method applied should be validated for the species and the trace metal(s) studied.

Destruction by the wet ashing chemical method takes place in teflon bombs at elevated temperatures and pressure (oven or microwave), or by the reflux method in the open atmosphere. The latter method is subject to possible contamination problems. Ashing of the tissue before dissolution is also an often applied technique (Bock, 1979). The enzymatic destruction and saponification of the tissue is relative new.

Once destruction is performed, many analytical techniques are available for the analysis of trace elements in biological samples (Stoeppler, 1991). The methods given for the analysis of trace metals in seawater (A-7.24) may, in principle, be applied to dissolved particulates.

(Instrumental) analytical methods for trace elements compare reasonably well. Although discrepancies may exist between laboratories, no systematic differences have been observed between the techniques proper.

To correct for seasonal differences, the results are expressed on the basis of dry weight (DW) or ash-free dry-weight (AFDW).

Various tissues, including mussel tissue, have been certified and are available as certified reference materials, which allows testing of analytical procedures to be performed (Cantillo, 1992).

**Sampling:**
Depending on their size, at least 25 individuals (preferably 50-100), identified to a species level and of even size (age) are collected. No metal tools are allowed and the use of plastic gloves will prevent contamination.

The organisms are collected in polythene plastic bags or wide mouth bottles.

The tissue of all organisms is homogenized and can be pooled into one composite sample. Sub-samples of the homogenate are analyzed.

Under no condition can (formalin) preserved samples be used for trace metal analysis.

**Sample volume:**
About 2-5 g (wet weight) of tissue homogenate will be required for analysis, but due to the sampling strategy more will usually be available.

**Sample treatment:**
After sampling, thorough homogenisation is essential. This can be performed using an Ultra-Turrax, Polytron blender, Tissumizer or suchlike. For trace metal work, the parts in contact with the tissue should not contain metal parts. Stainless steel blades and shaft may be replaced by titanium to prevent contamination. Homogenization should last 5 min.

**Storage:**
Wet tissue (homogenate) should be stored deep frozen (-18 °C or better). Completely dried/ashed tissue can be stored at ambient temperature.

**Analytical method(s):**
*Method #1a.* Destruction of wet tissue, followed by instrumental analysis.
*Method #1b.* Ashing of the tissue, dissolution in acid, dissolution, followed by instrumental analysis.

**Remarks:**
A special fume hood is required for working with perchloric acid. Refer to the safety procedures for working with acids.

**Method(s) references:**
Merian (1991)

**A-7.63**

| | |
|---|---|
| **Variable:** | **macrozoobenthos** |
| | **polycyclic aromatic hydrocarbons, PAHs** |
| **Unit:** | **mg/kg** |
| **Compartment:** | **sediment** |

**Introduction:**
PAHs have been determined successfully in 'Mussel Watch' type studies (*e.g.*
Boom, 1987; Murray *et al.*, 1991). Macrozoobenthos accumulate significant
amounts of PAHs from ambient waters and the concentrations of these compounds present in their tissues can provide information on the extent of contamination of estuarine and coastal waters (Boom, 1987). The international 'Mussel
Watch' type studies are examples of this type of approach (Cantillo, 1991;
O'Connor *et al.*, 1994; De Kock & Kramer, 1994). Mussels are considered to
have a low metabolic activity for trace organic pollutants, which is a requirement
for proper 'Mussel Watch' studies.
To avoid natural variation becoming dominant in the analysis, a relatively large
number of individuals need to be collected, then homogenised and pooled into
one composite sample. It is important that the individuals are identified to the
species level, as large differences in accumulation may occur, even between
apparently similar organisms. In order not to disturb the analysis, organisms
which contain much sediment in their stomachs (worms) need to depurate for
24 h in clean seawater before being homogenised. Mussels are usually not
depurated to minimize the loss of compounds of interest. Extraction of the trace
organics is required. After homogenisation, 'dissolution' by enzymatic methods,
or saponification, are useful techniques. When the method of extraction is optimized for the given trace organic pollutant and species, the available instrumental
analytical technique will determine the further procedures. Any method applied
should be validated for the species and the trace compounds studied.
To correct for seasonal differences, the results are expressed on the basis of dry
weight (DW) or ash-free dry-weight (AFDW). Although the concentrations are
sometimes based on lipid content, this may introduce additional problems for the
interpretation, as the variation in lipid content is often much higher than the
variation in pollutant concentration. When in doubt, the concentrations should be
reported based on both lipid content and DW.
Since many PAHs are considered toxic or carcinogenous, toxic equivalent factors
(TEFs) have been determined from a number of them (Nisbet & LaGoy, 1992).

**Sampling:**
Depending on their size, at least 25 individuals (preferably 50-100), identified to
a species level and of even size (age) are collected.

The organisms are collected in solvent pre-cleaned wide mouth glass bottles. The tissue of all organisms is homogenized and can be pooled into one composite sample. Sub-samples of the homogenate are analyzed.
Under no condition can (formalin) preserved samples be used for PAH analysis.

**Sample volume:**
About 20-50 g (wet weight) of tissue homogenate will be required for analysis.

**Sample treatment:**
After sampling, thorough homogenisation is essential. This can be performed using an Ultra-Turrax, Polytron blender, Tissumizer or suchlike. No plastic parts (except Teflon) should be in contact with the tissue. Stainless steel blades and shaft should be rinsed with an appropriate solvent (n-hexane, acetone). Homogenization should last 5 min.

**Storage:**
Wet tissue (homogenate) should be stored deep frozen (-18 °C or lower).

**Analytical method(s):**
*Method #1a.* The sample is homogenized and the tissue disrupted by saponification or by enzymatic treatment, after which it is submitted to solvent extraction using n-hexane.
The extract is reduced in volume and cleaned by column chromatography using a suitable sorbent such as florisil, silica gel or aluminum oxide. The cleaned extract is concentrated and analyzed using reversed phase high performance liquid chromatography (RP-HPLC) with fluorescence and UV-absorption detection. Quantitation is based on calibration using external standards.
*Methods #1b.* Alternatively, the sample may be submitted to extraction without prior saponification or enzymatic treatment. This may require proper selection of the extraction solvent (mixture).
*Methods #2a,b.* Instead of reversed phase high performance liquid chromatography with fluorescence and UV-absorption detection, capillary gas chromatography with mass-spectrometric (GC/MS) detection in the selected-ion monitoring (SIM) mode may be used. Identification of individual PAHs is based on combined retention data, the specific ions detected and (optionally) ion-abundance ratios.
*Methods #3a,b.* Before extraction, a mixture of $^{13}C_{12}$ labelled PAHs is added to the sample as an internal standard (isotope dilution). The sample is treated by one of the Methods #1a-c, described above, and analyzed using capillary gas-chromatography with mass-spectrometric detection in the selected-ion monitoring mode. Identification of individual PAHs is based on a combination of retention data, the specific ions detected and ion-abundance ratios. Quantitation is based on the comparison of responses with those of corresponding internal standards.

**Remarks:**

**Method(s) references:**
Neff (1979); Lee *et al.* (1981); UNEP (1990); Ehrhardt *et al.* (1991)

## A-7.64

**Variable:**     **macro-zoobenthos**

               **polychlorobiphenyls, PCBs**

**Unit:**        **µg/kg**

**Compartment:**  **sediment**


**Introduction:**
Macrozoobenthos accumulate significant amounts of trace organic pollutants, including PCBs, from ambient waters and the concentrations of these compounds present in their tissues can provide information on the extent of contamination of estuarine and coastal waters. The international 'Mussel Watch' type studies are examples of this type of approach (Cantillo, 1991; O'Connor et al., 1994; De Kock & Kramer, 1994).

Bivalves lend themselves very well to the approach of biomonitoring, and often used representatives are mussels (*Mytilus* spp., *Perna* spp.), oysters (*Ostrea* spp., *Crassostrea* spp.) and clams. They are considered to have a low metabolic activity for trace organic pollutants. Other benthic organisms have also been used in PCB monitoring studies *e.g.* worms (*Arenicola marina, Nereis* spp.), crustacea (shrimps, barnacles, crabs), echinoderms, etc.

To avoid natural variation becoming dominant in the analysis, a relatively large number of individuals need to be collected, then homogenised and pooled into one composite sample. It is important to identify individuals to the species level, as differences may occur, even between apparently similar organisms. In order not to disturb the analysis, organisms which contain much sediment in their stomachs (worms) need to depurate for 24 h in clean seawater before being homogenised. Mussels are usually not depurated to minimize for the loss of compounds of interest.

Extraction of the trace organics is required. After homogenisation 'dissolution' by enzymatic methods, or saponification, are useful techniques. When the method of extraction is optimized for the given trace organic pollutant and species, the available instrumental analytical technique will determine the further procedures. Any method applied should be validated for the species and the set of PCBs studied.

To correct for seasonal differences, the results are expressed on the basis of dry weight (DW) or ash-free dry-weight (AFDW). Although the concentrations are sometimes based on lipid content, this may introduce additional problems for the interpretation, as the variation in lipid content is often much higher than the variation in PCB concentration. When in doubt, the concentrations should be reported based on both lipid content and DW.

In order to relate the toxicity to a standard (dioxin), toxicity equivalent factors (TEFs) have been determined (Safe, 1990).

**Sampling:**
Depending on their size, at least 25 individuals (preferably is 50-100), identified to a species level and of even size (age) are collected.
The organisms are collected in solvent pre-cleaned wide mouth glass bottles.
The tissue of all organisms is homogenized and can be pooled into one composite sample. Sub-samples of the homogenate are analyzed.
Under no condition can (formalin) preserved samples be used for PCB analysis.

**Sample volume:**
About 20-50 g (wet weight) of tissue homogenate will be required for analysis.

**Sample treatment:**
After sampling, thorough homogenisation is essential. This can be performed using an Ultra-Turrax, Polytron blender, Tissumizer or suchlike. No plastic parts (except Teflon) should be in contact with the tissue. Stainless steel blades and shaft should be rinsed with an appropriate solvent (n-hexane, acetone). Homogenization should last 5 min.

**Storage:**
Wet tissue (homogenate) should be stored deep frozen (-18 °C or better).

**Analytical method(s):**
*Method #1a.* The sample is homogenized and the tissue disrupted by saponification or by enzymatic treatment, after which it is submitted to solvent extraction using n-hexane, petroleum ether, or binary or tertiary solvent mixtures. Acetone, if used, is removed by back-extraction with water.
The extract is reduced in volume and cleaned by column chromatography using a suitable sorbent such as florisil, silica gel or aluminum oxide.
The cleaned extract is concentrated and analyzed by capillary gas chromatography (GC) with electron capture detection (ECD). Identification of individual PCBs is based on retention data, preferably using two GC columns with different polarity. Quantitation is based on calibration using external standards.
*Methods #1b.* Alternatively, the sample may be submitted to extraction without prior saponification or enzymatic treatment. This may require proper selection of the extraction solvent (mixture).
*Methods #2a,b.* Instead of capillary gas chromatography with electron capture detection, capillary gas chromatography with mass-spectrometric detection in the selected-ion monitoring mode may be used.
Identification of individual PCBs is based on combined retention data, specific ions detected and (optionally) ion-abundance ratios.
*Methods #3a,b.* Before extraction, a mixture of $^{13}C_{12}$ labelled PCBs is added to the sample as an internal standard (isotope dilution).
The sample is treated by one of the Methods #1a or #1b described above, and analyzed using capillary gas chromatography with mass-spectrometric detection in the selected-ion monitoring mode.

Identification of individual PCB is based on a combination of retention data, the specific ions detected and ion-abundance ratios. Quantitation is based on the comparison of responses with those of corresponding internal standards.

**Remarks:**

**Method(s) references:**
Erickson (1986); Waid (1986, 1987)

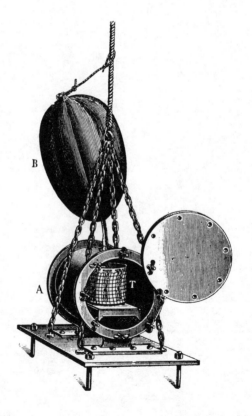

*The past:*
*Registrating thermometer of Regnard (From: Richard, 1907)*

# 8. DATABASE

One of the necessities for introduction of information from different sources and compiled by various scientists into a database is the uniformity of the data. The use of the same units is prerequisite for uniformity, but also methods of collection (sampling, analysis) should ideally be the same. It is obvious that this can never be completely achieved (CCME, 1993).

The compilation of a data base comprising all the measurements carried out in the project was one of the major tasks of the JEEP92 project. This database was to be fully computerized. In order to deal with streams of data originating from at least as many sources as there are participants in the project, a data input and editing program had to be developed and made available to the participants.

One of the aim of this manual was to guide the users for the JEEP92 project in the presentation of the proper units, and in the identification of sampling and analytical Methods. Those Methods, which are described in the sections on sampling and analysis (chapters 6 and 7) were also included in the database. Data collected were to be stored in the database together with coded information about the method of collection and the method of analysis. This ensures that, if required, data collected with identical methods may be separated from those data which were gathered by techniques that do not compare well. This is not only true for physico/chemical variables like salinity, temperature, etc. of which uniformity is rather simple to perform, but also for variables that are the result of different methods, *e.g.* biomass calculated from size, chlorophyll or measured organic carbon content. Also scientific names of biological species (or even families) should be introduced in a uniform way. In the database the change of scientific names, because new taxonomic information has become available, should be a simple operation.

For the JEEP92 project, a database format has been specially designed for the collection (and interpretation) of the data from various estuaries. This computer program, the Data Entry and Editing Program (DEEP-92), was published in a separate report (De Winter, 1992a).

This data entry program has been especially developed for the entry of data of the JEEP92 programme, and was written using DBASE IV. It will function on any computer platform that is able to use this database program.

For the JEEP92 project, which operated in western European estuaries, rules regarding defined species names have been put forward to facilitate data entry which involve estuarine species names. The names of the genus or species and their spelling are preferably to be selected here from the "Directory of the British marine fauna and flora" (Howson, 1987). This list is available both in printed format as well as on floppy disk. Some species or genera will not be available from this list. In that case the database manager will decide how to proceed. For other areas equivalent standard works should be available.

It is realized that probably none of the participants of a given project can be a specialist in the identification of all individual species of all estuarine taxa. It is strongly advised, however, to identify the organisms at least to the gen-era level, thus for example *Arenicola* spp.

# 9. GLOSSARY

| | |
|---|---|
| AAS | atomic absorption spectrometry |
| ADPCSV | adsorptive differential pulse CSV |
| AFDW | ash free dry weight |
| AFS | atomic fluorescence spectrometry |
| APDC | ammonium pyrolidine-1-dithiocarbonic acid |
| CHN | carbon hydrogen nitrogen |
| COST | Cooperation européenne dans le domaine de la recherche Scientifique et Technique |
| CSV | cathodic stripping voltammetry |
| CTD | conductivity temperature depth |
| DDDC | di-ethylammonium-N,N-di-ethyldithio-carbamate |
| DNA | deoxyribo nucleic acid |
| DOC | dissolved organic carbon |
| DON | dissolved organic nitrogen |
| DPASV | differential pulse anodic stripping voltammetry |
| DW | dry weight |
| ETAAS | electro-thermal AAS |
| GC-MS | gas chromatography mass spectrometry |
| HPLC | high performance liquid chromatography |

| | |
|---|---|
| IAPSO | International Association for the Physical Sciences of the Ocean |
| ICP-AES | inductively coupled plasma atomic emission spectrometry |
| ICP-MS | inductively coupled plasma mass spectrometry |
| MAST | Marine Science and Technology |
| MPN | most probable number |
| NAA | neutron activation analysis |
| nm | nano meter |
| PE | poly ethylene |
| POC | particulate organic carbon |
| PON | particulate organic nitrogen |
| POP | particulate organic phosphorus |
| PVC | poly vinyl chloride |
| SD | Secchi depth |
| SED | spheric equivalent diameter |
| SPM | suspended particulate matter |
| ST | salinity - temperature |
| TLC | thin layer chromatography |
| TTI | tritiated thymidine incorporation |
| TXRF | total reflection X-ray fluorescence |
| μm | micro meter |
| UNESCO | United Nations Educational, Scientific and Cultural Organisation |
| UV | ultra violet |
| ZETAAS | ETAAS with Zeeman background correction |

# 10. REFERENCES

Abaychi, J.K. & J.P. Riley, 1979. The determination of phytoplankton pigments by high-performance liquid chromatography. Anal. Chim. Acta, 107: 1-11

Ackermann, F., H. Bergmann & W. Schleichert, 1983. Monitoring of heavy metals in coastal and estuarine sediments. A question of grain-size: < 20 µm versus < 60 µm. Environm. Technol. Lett. 4: 317-328

Aminot, A. &. M. Chaussepied, 1983. Manuel des analyses chimiques en milieu marin. Centre national pour l'exploration des océans (CNEXO). Brest, pp. 395

Anon., 1990. Guidelines for the sampling and analysis of trace metals in seawater under the Joint Monitoring Programme (JMP). Annex to the recommendations at the fifteenth meeting of JMG, Lisbon, 1990.

APHA, AWWA & WPCF, 1985. Standard methods for the examination of water and wastewater, 16th edition. American Public Health Association, American Water Works Association, Water Pollution Control Federation, Washington, pp. 1268

Aston, S.R., 1981. Estuarine chemistry. In: Chemical oceanography, 2nd ed. Vol. 7, J.P. Riley & R. Chester (eds). Academic Press, London, pp. 361-440

Bachelet, G., 1990. The choice of a sieving mesh size in the quantitative assessment of marine macrobenthos: a necessary compromise between aims and constraints. Mar. Environ. Res. 30: 21-35`

Ballschmiter, K. & M. Zell, 1980. Analysis of polychlorinated biphenyl (PCB) by glass capillary gas chromatography. Composition of technical Aroclor and Clophen PCB mixtures. Fres. Z. Anal. Chem. 302: 20-31

Baker, J.M. & W.J. Wolff (eds), 1987. Biological surveys of estuaries and coasts. EBSA Handbook, Cambridge Univ. Press, Cambridge, pp. 449

Barnet, B.E., 1979. Sorting benthic samples. Mar. Poll. Bull. 10: 241-242

Barth, W.F., 1952. Theoretical petrology. Wiley, New York

Baskerville, G.L., 1972. Use of logarithmic regression in the estimation of plant biomass. Can. J. Forrestry, 2: 49-53

Batley, G.E. (ed), 1989. Trace element speciation: analytical methods and problems. CRC Press, Boca Raton FL, pp. 350

Baudo, R., 1990. Sediment sampling, mapping and data analysis. In: Sediments: chemistry and toxicity of in-place pollutants; R. Baudo, J. Giesy & H. Muntau (eds). Lewis Publ., Boca Raton, pp. 15-60

Beers, J.R., 1978. Sampling techniques: pump sampling. In: Phytoplankton manual. A. Sournia (ed). Unesco, Paris, pp. 41-49

Berglund, F. & B. Sörbo, 1960. Turbidimetric analysis of inorganic sulfate in serum, plasma and urine. Scandinav. J. Clin. and Lab. Investig. 12: 147-153

Berman, S.S., R.E. Sturgeon, J.A.H. Desaulniers & A.P. Mykytiuk, 1983. Preparation of the seawater reference material for trace metals, NASS-1. Mar. Pollut. Bull. 14: 69-73.

Berman, S.S. & Yeats, P.A., 1987. Sampling of seawater for trace metals. CRC Crit. Rev. in Anal. Chem. 16: 1-14

Bewers, J.M. & H.L. Windom, 1982. Comparison of sampling devices for trace metal determinations in seawater. Mar. Chem. 11: 71-86.

Bewers, J.M., P.A. Yeats, S. Westerlund, B. Magnusson, D. Schmidt, H. Zehle, S.S. Berman, A. Mykytiuk, J.C. Duinker, R.F. Nolting, R.G. Smith & H.L. Windom, 1985. An intercomparison of seawater filtration procedures. Mar. Poll. Bull. 16: 277-281

Bhattacharya, C.G., 1967. A simple method of resolution of a distribution into Gaussian components. Biometrics, 23: 115-135

Bhaud, M., 1979. Description et utilisation d'un engin de récolte à proximité du sédiment. Bull. Ecol. 10: 15-27

Billington, N., 1991. A comparison of three methods of measuring phytoplankton biomass on a daily and seasonal basis. Hydrobiol. 226: 1-15

Bock, R., 1979. A handbook of decomposition methods in analytical chemistry. Int. Textbook Company, pp. 444

Boom, M.M., 1987. The determination of polycyclic aromatic hydrocarbons in indigenous and transplanted mussels (*Mytilys edulis* L.) along the Dutch coast. Intern. J. Environ. Anal. Chem. 13: 251-261

Bouwman, L.A., 1987. Meiofauna. In: Biological surveys of estuaries and coasts, J.M. Baker, & W.J. Wolff (eds). EBSA Handbook, Cambridge Univ. Press, Cambridge, pp. 140-156

Brennan, B., 1989. Intercalibration report from working group 3/4. SHANELBE project, Trinity College, Dublin, pp. 4

Brockmann, U.H. & G. Hentzschel, 1983. Samplers for enclosed stratified water columns. Mar. Ecol. Prog. Ser. 14: 107-109

Bros & Cowell, 1987. A technique for optimizing sample size (replication). J. Exp. Mar. Biol. Ecol. 114: 63-71

Brunel, P., M. Besner, D. Messier, L. Poirier, D. Granger & M. Weinstein, 1978. Le traîneau Macer-Giroq: appareil amélioré pour l'échantillonage quantitatif étagé de la petite faune nageuse au voisinage du fond. Intern. Rev. gesamten Hydrobiol. 63: 815-829

Buchanan, J.B., 1984. Sediment analysis. In: Methods for the study of marine benthos. N.A. Holme & A.D. McIntyre (eds). 2nd ed. Blackwell, Oxford, 41-65

Burton, G.A., 1992. Sediment collection and processing: factors affecting realism. In: Sediment toxicity assessment; G.A. Burton (ed), Lewis Publ., Boca Raton, pp. 37-66

Burton, J.D. & P.S. Liss (eds), 1976. Estuarine chemistry. Academic Press, London, pp. 299

Cairns, J. & J.R. Pratt, 1986. Developing a sampling strategy. Rationale for sampling and interpretation of ecological data in the assessment of fresh-water ecosystems. In: B.G. Isom (ed). report ASTM STP 894. Philadelphia, 168-186

Cadée, G.C. & J. Hegeman, 1974. Primary production of the benthic microflora living on tidal flats in the Dutch Wadden Sea. Neth. J. Sea Res. 8: 260-291

Cadée, G.C. & J. Hegeman, 1977. Distribution of primary production of the benthic microflora and accumulation of organic matter on a tidal flat area, Balgzand, Dutch Wadden Sea. Neth. J. Sea Res. 11: 24-41

Cantillo, A.Y., 1991. Mussel Watch worldwide literature survey - 1991. NOAA Technical memorandum NOS ORCA 63, Rockville, MD, pp. 142

Cantillo, A.Y., 1992. Standard and reference materials for marine science (3rd Ed). NOAA National status and trends program, U.S. Dept of Commerce, Rockville, MD, USA, pp. 577

Cassie, R.M., 1954. Some uses of probability paper in the analysis of size frequency distributions. Aust. J. Mar. Freshw. Res. 5: 513-522

Caspers, H., 1959. Die Einteilung der Brackwasser-Regionen in einem Aestuar. Archo. Oceanogr. Limnol. (Suppl.), 11: 153-160

Cauwet, G., 1981. Non-living particulate matter. In: Marine organic chemistry, E.K. Duursma & R. Dawson (eds). Elsevier, Amsterdam, pp. 71-89

CCME, 1993. Guidance manual on sampling, analysis, and data management for contaminated sites. Volume I: Main Report; Volume II: Analytical method summaries. Canadian Council of Ministers of the Environment, The National Contaminated Sites Remediation Program. Report CCME EPC-NCS66E, Winnipeg, Manitoba, pp. 171

Clarke, J.U., V.A. McFarland & B.D. Pierce, 1989. Preliminary recommendeations for a congener-specific PCB analysis in regulatory evaluation of dredged material. Misc. paper D-89-2, US Army Engineer Waterways Experiment Station, Vicksburg, MS, pp. 33

Coleman, N., 1980. More on sorting benthic samples. Mar. Poll. Bull. 11: 150-152

Colijn, F. & K.S. Dijkema, 1981. Species composition of benthic diatoms and distribution of Chlorophyll a on an intertidal flat in the Dutch Wadden Sea. Mar. Ecol. Progr. Ser. 4: 9-21

Colijn, F. & V.N. De Jonge, 1984. Primary production of micro-phytobenthos in the Ems-Dollard estuary. Mar. Ecol. Prog. Ser. 14: 185-196

Copin-Montegut, C. & G. Copin-Montegut, 1973. Comparison between two processes of determination of particulate organic carbon in sea water. Mar. Chem. 1: 151-156

Craib, J.S., 1965. A sampler for short undisturbed cores. J. Cons. Perm. Int. Expl. Mer, 30: 346-353

Crisp, D.J., 1984. Energy flow measurements. In: Methods for the study of marine benthos. N.A. Holme & A.D. McIntyre (eds). Blackwell Sci. Publ., Oxford, pp. 284-372

Dankers, N. & R. Laane, 1983. A comparison of wet oxidation and loss on ignition of organic material in suspended matter. Environm. Technol. Lett. 4: 283-290

Danielson, L.-G., B. Magnusson, S. Westerlund & K. Zhang, 1982. Trace metal determinations in estuarine waters by electrothermal AAS after extraction in dithiocarbamate complexes in freon. Anal. Chim. Acta 144: 183-188

Dauvin, J.C. & J.C. Lorgère, 1989. Modifications de traîneau Macer-Giroq pour l'amélioration de l'échantillonage quantitatif étagé de la faune suprabenthique. J. Rech. Océanogr. 14: 65-67

Dawson, R. & G. Liebezeit, 1981. The analytical methods for the characterization of organics in seawater. In: Marine organic chemistry, E.K. Duursma & R. Dawson (eds). Elsevier, Amsterdam, pp. 445-496

Day, J.H. (ed), 1981. Estuarine ecology with particular reference to southern Africa. Balkema, Rotterdam, pp. 441

Day, J.W., C.A.S. Hall, W.M. Kemp & A. Yanes-Arancibia (eds), 1989. Estuarine ecology. Wiley, New York, pp. 558

De Jonge, V.N., 1979. Quantitative separation of benthic diatoms from sediments using density gradient centrifugation in colloidal silica Ludox-TM. Mar. Biol. 51: 267-278

De Jonge, V.N., 1980. Fluctuations in the organic carbon to chlorophyll-a ratios for estuarine benthic diatom populations. Mar. Ecol. Progr. Ser. 2: 345-353

De Jonge, V.N., 1992. Physical processes and dynamics of microphytobenthos in the Ems estuary (the Netherlands). PhD thesis, Univ. Groningen, pp. 176

De Jonge, V.N. & L.A. Bouwman, 1977. A simple density separation technique for quantitative isolation of meiobenthos using using the colloidal silica Ludox-TM. Mar. Biol. 42: 143-148

De Kock, W.C. & K.J.M. Kramer, 1994. Active biomonitoring (ABM) by translocation of bivalve molluscs. In: Biomonitoring of coastal waters and estuaries, K.J.M. Kramer (ed). CRC Press, Baton Rouge, pp. 51-84

De Winter, W.P., 1992a. DEEP-92. The Data Entry and Editing Program of JEEP-92. User manual. JEEP92 report, NIOO-CEMO, Yerseke.

Demers, S. (ed), 1991. Particle analysis in oceanography. NATO ASI series, G: Ecological sciences Vol. 27. Springer, Berlin

DiToro, D.M., C.S. Zarba, D.J. Hansen, W.J. Berry, R.C. Swartz, C.E. Cowan, S.P. Pavlou, H.E. Allen, N.A. Thomas & P.R. Paquin, 1991. Technical basis for establishing sediment quality criteria for nonionic organic chemicals using equilibrium partitioning. Environ. Toxicol. Chem. 10: 1541-1583

Downing, J.A., 1979. Aggregation, transformation, and the design of benthos sampling programs. Can. J. Fish. aquat. Sci. 36: 1454-1463,

Downing, J.A., 1989. Precision of the mean and the design of benthic sampling programmes: caution revised. Mar. Biol. 103: 231-234

Duinker, J.C., R.F. Nolting & H.A. Van der Sloot, 1979. The determination of suspended metals in coastal waters by different sampling and processing techniques (filtration, centrifugation). Neth. J. Sea Res. 13: 282-297

Duinker, J.C. & M.T.J. Hillebrand, 1983. Determination of selected organochlorines in seawater. In: Methods of seawater analysis; K. Grasshoff, M. Ehrhardt & K. Kremling (eds). Verlag Chemie, Weinheim, pp. 290-309

Duinker, J.C., A.H. Knap, K.C. Binkley, G.H. van Dam, A. Darrel Rew & M.T.J. Hillebrand, 1988. Method to represent the qualitative and quantitative characteristics of PCB mixtures. Marine mammal tissues and commercial mixtures as examples. Mar. Poll. Bull. 19: 74-79

Durst, R.A., 1979. Container materials for the preservation of trace substances in environmental specimens. Monitoring environmental materials and specimen banking, N.-P. Luepke (ed). Martinus Nijhoff, The Hague, pp. 198-202.

Dustan, P. & J.L. Pinckney, 1989. Tidally induced estuarine phytoplankton patchiness. Limnol. Oceanogr. 34: 410-419

Duursma, E.K. & R. Dawson (eds), 1981. Marine organic chemistry. Elsevier, Amsterdam, pp. 521

Dyer, K.R., 1979. Estuarine hydrography and sedimentation. Cambridge University Press, Cambridge

Dyer, K.R., 1991. Circulation and mixing in stratified estuaries. Mar. Chem. 32: 111-120

Eaton, J.W. & B. Moss, 1966. The estimation of numbers and pigment content in epipelic algal populations. Limnol. Oceanogr. 11: 584-595

Eberlein, K. & K.D. Hammer, 1980. Automatic determination of total carbohydrates in seawater. Fres. Z. Anal. Chem. 301: 17-19

Eberlein, K. & L. Schütt, 1986. Automatic methods for the determination of total dissolved and particulate carbohydrates in the marine environment. Fres. Z. Anal. Chem. 323: 47-49

Eberlein, K. & G. Kattner, 1987. Automatic method for the determination of ortho-phosphate and total dissolved phosphorus in the marine environment. Fres. Z. Anal. Chem. 326: 354-357

Ehrhardt, M., J. Klungsøyr & R.J. Law., 1991. Hydrocarbons: Review of methods for analysis in seawater, biota and sediments. Techniques in marine environmental sciences No. 2, ICES, Copenhagen, pp. 47

Eisma, D., 1986. Flocculation and de-flocculation of suspended matter in estuaries. Neth. J. Sea Res. 20: 183-199

Eisma, D., T. Schuhmacher, H. Boekel, J. van Heerwaarden, H. Franken, M. Laan, A. Vaars, F. Eijgenraam & J. Kalf, 1990. A camera and image-analysis system for in situ observation of flocks in natural waters. Neth. J. Sea Res. 27: 43-56

Eleftheriou, A. & N.A. Holme, 1984. Macrofauna techniques. In: Methods for the study of marine benthos. N.A. Holme & A.D. McIntyre (eds). 2nd ed. Blackwell, Oxford, pp. 140-216

Elliot, J.M., 1977. Some methods for the statistical analysis of samples of invertebrates, 2nd ed. Scientific publications 25, Ambleside, Freshwater Biological Association

Elliott, J.M., P.A. Tullett & J.A. Elliott, 1993. A new bibliography of samplers for freshwater benthic invertebrates. NERC Freshwater Biological Association, occasional publication No. 30, Ambleside, pp. 92

EPA, 1992. Methods for the determination of metals in environmental samples. C.K. Smoley, Boca Raton Fl. pp. 339

Erickson, M.D., 1986. Analytical chemistry of PCB's. Ann Arbor Science, Ann Arbor

Essink, K. & J.J. Beukema, 1991. Long-term changes in intertidal and shallow-subtidal sedimentary zoobenthos. Review of work carried out within the framework of COST 647. In: Space and time series data in coastal benthic ecology, B.F. Keegan (ed), Commission of the European Communities, Brussels, pp. 43-64

Essink, K. & H.L. Kleef, 1991. Comparison of sampling strategies for trend monitoring of intertidal sedimentary macrozoobenthos. In: B.F. Keegan (ed), Commission of the European Communities, DG XII, Environmental Research Programme. COST 647 Coastal Benthic Ecology, Activity report 1988-1991, Brussels, pp. 287-294

Fairbridge, R.W., 1980. The estuary: its definition and geo-dynamic cycle. In: Chemistry and biochemistry of estuaries, E. Olausson & I. Cato (eds). Wiley, Chichester, pp. 1-37

Fanger, H.U., H. Kuhn, W. Michaelis, A. Müller & R. Riethmüller, 1986. Investigation of material transport and load in tidal rivers. Wat. Sci. Technol. 18: 101-110

Fanger, H.U., J. Kappenberg, H. Kuhn, U. Maixner & D. Milferstaedt, 1990. The hydrographic measuring system HYDRA. In: Estuarine water quality management, W. Michaelis (ed). Springer, Berlin, pp. 211-216

Feller, R.J. & R.M. Warwick, 1988. Energetics. In: Introduction to the study of meiofauna. R.P. Higgins & H. Thiel (eds). Smithsonian Inst. Press, Washington DC, pp. 181-196

Fleeger, J.W., D. Thistle & H. Thiel, 1988. Sampling equipment. In: Introduction to the study of meiofauna. R.P. Higgins & H. Thiel (eds). Smithsonian Inst. Press, Washington DC, pp. 115-125

Förstner, U. & W. Salomons, 1980. Trace analysis on polluted sediments. I. Assessment of sources and intensities. Environm. Technol. Lett. 1: 495-505

Förstner, U., 1989. Contaminated sediments. Lectures on environmental aspects of particle-associated chemicals in aquatic systems. Lecture notes in earth sciences, 21. Springer, Berlin, pp. 157

Freitas, M.C., R. Cornelis, F. Decorte & L. Mees, 1993. Sample preparation of aquatic macrophytes for analysis of minor-elements and trace-elements. Sci. Total Environ. 130: 109-120

Fuhrman, J.A. & F. Azam, 1982. Thymidine incorporation as a measure of heterotrophic bacterioplankton production in marine surface waters: evaluation and field results. Mar. Biol. 66: 109-120

Garrasi, C., E.T. Degens & K. Mopper, 1979. Amino acid composition of seawater Mar. Chem. 8: 71-85

Gershey, R.M., M.D. MacKinnon & P.J. LeB. Williams, 1979. Comparison of three oxidation methods used for the analysis of dissolved organic carbon in seawater. Mar. Chem. 7: 289-306

Gibbs, R.J., 1982. Floc stability during Coulter counter size analysis. J. Sed. Petrol. 52: 657-660

Gibbs, R.J. & L.N. Konwar, 1983. Effect of pipetting on mineral flocs. Environm. Sci. Technol. 16: 119-121

Giere, O., A. Eleftheriou & D.J. Murison, 1988. Abiotic factors. In: Introduction to the study of meiofauna. R.P. Higgins & H. Thiel (eds). Smithsonian Inst. Press, Washington DC, pp. 61-78

Gieskes, W.W. & G.W. Kraay, 1983. Unknown chlorophyll a derivatives in the North Sea and tropical Atlantic Ocean revealed by HPLC analysis. Limnol. Oceanogr. 28: 757-765

Gomez-Parra, A., J.M. Fonja & D. Cantero, 1987. A new device for sampling waters in shallow ecosystems. Wat. Res. 21: 1437-1443

Gorsline, D.S., 1984. A review of fine-grained sediment. Origins, characteristics, transport and deposition. In: Fine grained sediments: deep water processes and facies, D.A.V. Stow & D.J.M. Piper (eds). Blackwell, Oxford, pp. 17-34

Grasshoff, K., M. Ehrhardt & K. Kremling (eds), 1983. Methods of seawater analysis. 2nd ed. Verlag Chemie, Weinheim

Green, J., 1968. The biology of estuarine animals. Sidgwick & Jackson, London, pp. 401

Green, R.H., 1979. Sampling design and statistical methods for environmental biologists. Wiley, New York, pp. 253

Greiser, N. & A. Faubel, 1988. Biotic factors. In: Introduction to the study of meiofauna. R.P. Higgins & H. Thiel (eds). Smithsonian Inst. Press, Washington DC, pp. 79-114

Haake, B., V. Ittekot, R.R. Nair, V. Ramaswamy & S. Honjo, 1992. Fluxes of amino acids and hexosamino to the deep Arabian Sea. Mar. Chem. 40: 291-314

Hagström, A., U. Larsson, P. Hörstedt & S. Normark, 1979. Frequency of dividing cells, a new approach to the determination of bacterial growth rates in aquatic environments. Appl. Environ. Microbiol. 37: 805-812

Hamerlynck, O. & J. Mees, 1991. Temporal and spatial structure in the hyperbenthic community of a shallow coastal area and its relation to environmental variables. Oceanol. Acta, 11: 205-211

Hammer, K.D. & K. Nagel, 1986. An automated fluorescence assay for sub-nanogram quantities of protein in the presence of interfering material. Anal. Biochem. 155: 308-314

Hammer, K.D. & S. Luck, 1987. An automated fluorescence assay for dissolved amino acids from marine waters in the presence of interfering material. Fres. Z. Anal. Chem. 327: 518-520

Harding, J.P., 1949. The use of probability paper for the graphical analysis of poly-modal frequency distributions. J. Mar. biol. ass. UK, 28: 141-153

Hartley, J.P. & B. Dicks, 1987. Macrofauna of subtidal sediments using remote sampling. In: Biological surveys of estuaries and coasts, J.M. Baker, & W.J. Wolff (eds). EBSA Handbook, Cambridge Univ. Press, Cambridge, pp. 106-130

Hartley, J.P., B. Dicks & W.J. Wolff, 1987. Processing sediment macrofauna samples. In: Biological surveys of estuaries and coasts, J.M. Baker, & W.J. Wolff (eds). EBSA Handbook, Cambridge Univ. Press, Cambridge, pp. 131-139

Head, P.C., 1985a. Salinity, dissolved oxygen and nutrients. In: Practical estuarine chemistry, P.C. Head (ed), Cambridge Univ. Press, Cambridge, pp. 94-125

Head, P.C. (ed), 1985b. Practical estuarine chemistry. Cambridge Univ. Press, Cambridge, pp. 337

Heinle, D.R. & D.A. Flemer, 1975. Carbon requirements of a population of the estuarine copepod Eurytemora affinis. Mar. Biol. 31: 235-247

Hermans, J.H. Smedes, F. Hofstraat, J.W. Cofino, W.P., 1992. A method for estimation of chlorinated biphenyls in surface waters - influence of sampling method on analytical results. Environ. Sci. Technol. 26: 2028-2035

Higgins, R.P. & H. Thiel, 1988. Introduction to the study of meiofauna. Smithsonian Institution Press, Washington DC, pp. 488

Hillebrand, M.T.J & R.F. Nolting, 1987. Sampling procedures for organochlorines and trace metals in open ocean waters. Trends Anal. Chem. 6: 74-77

HMSO, 1980. Dissolve oxygen in natural and waste waters. Methods for the examination of waters and associated materials. London, Her Majesty's Stationary Office

Hobbie, J.E., R.J. Daley & S. Jasper, 1977. Use of Nuclepore filters for counting bacteria by fluorescence microscopy. Appl. Environ. Microbiol. 33: 1225-1228

Holme, N.A., 1964. Methods of sampling the benthos. Adv. Mar. Biol. 2: 171-260,

Holme, N.A. & A.D. McIntyre (eds), 1984. Methods for the study of marine benthos. Blackwell Sci. Publ., Oxford, 387

Holm-Hansen, O., C.J. Lorenzen, R.W. Holmes & J.D.H. Strickland, 1965. Fluorimetric determination of chlorophyll. J. Cons. Perm. Int. Explor. Mer, 30: 3-15

Honda, S., E. Akao, S. Suzuki, M. Kakehi & J. Nakamura, 1989. High performance liquid chromatography of reducing carbohydrates as strongly ultraviolet absorbing and electrochemically sensitive 1-phenyl-3-methyl-5-pyrazolone derivatives. Anal. Biochem. 180: 351-357

Horowitz, A.J., K.A. Elrick & M.R. Colberg, 1992. The effect of membrane filtration artifacts on dissolved trace element concentrations. Wat. Res. 26: 753-763.

Howson, C.M., 1987. Directory of the British marine fauna and flora. Marine Conservation Society (also available on diskette)

Ittekot, V., W.G. Deuser & E.T. Degens, 1984. Seasonality in fluxes of sugars, amino acids and amino sugars to the deep ocean: Sargasso Sea. Deep Sea Res. 31: 1057-1069

Jeffrey, D.W., J.G. Wilson, C.R. Harris & D.L. Tomlinson, 1985. A manual for the evaluation of estuarine quality, 2nd ed. Univ. of Dublin, pp. 161

Jeffreys, S.W. & G.F. Humphrey, 1975. New spectrophotometric equations for determining chlorophylls a, b and c and $c_1$ in higher plants, algae and natural phytoplankton. Biochemie u. Physiologie d. Planzen, 167: 191-194

Johnson, C.M. & H. Nishita, 1952. Microestimation of sulphur in plant materials, soils, and irrigation waters. Anal. Chem. 24: 736-742

Johnson, K.M. & J.McN. Sieburth, 1977. Dissolved carbohydrates in seawater. I. A precise spectrophotometric analysis for monosaccharides. Mar. Chem. 5: 1-13

Jørgensen, S.E., S.N. Nielsen & L.A. Jørgensen, 1991. Handbook of ecological parameters and ecotoxicology. Elsevier, Amsterdam, pp. 1263

Kalle, K., 1937. Nahrstoff Untersuchungen als Hydrographisches Hilfsmittel zur Unterscheidung von Wasserkurpern. Ann. Hydrogr. Berlin, 65: 276-282

Kalle, K., 1963. Uber das Verhalten und der Herkunft der in den Gewassern und der Atmosphare vorhandenen himmelblauen Fluorescence. Dt. Hydrogr. Z. 16: 153-166

Kattner, G. & U.H. Brockmann, 1980. Semi-automated methods for the determination of particulate phosphorus in the marine environment. Fres. Z. Anal. Chem. 301: 14-16

Keith, L.H., W. Mueller & D.L. Smith (eds), 1991. Compilation of E.P.A.'s sampling and analysis methods. Lewis Publ. Boca Raton Fl., pp. 803

Kennedy, V.S. (ed), 1984. The estuary as a filter. Academic Press, New York, pp. 510

Ketchum, B.H. (ed), 1983. Ecosystems of the world 26: Estuaries and enclosed seas. Elsevier, Amsterdam, pp. 500

Kimmerer, W.J. & A.D. McKinnon, 1986. Glutaraldehyde fixation to maintain biomass of preserved plankton. J. Plankton Res. 8: 1003-1008

Kirkwood, D., A. Aminot & M. Perttilä, 1991. Report on the results of the ICES fourth intercomparison exercise for nutrients in seawater. ICES report 174, Copenhagen, pp. 83

Klamer, J.C., W.J.M. Hegeman & F. Smedes, 1990. Comparison of grain size correction procedures for organic micropollutants and heavy metals in marine sediments. Hydrobiol. 208: 213-220

Koopmann, C., J. Faller, K.-H. van Bernem, A. Prange & A. Müller, 1993. Schadstoffkartierung in Sedimenten de deutschen Wattenmeeres Juni 1989 - Juni 1992. UBA-FuE Report 109 03 377, GKSS, Geesthacht

Kramer, C.J.M. & Duinker, J.C. (Eds), 1984. Complexation of trace metals in natural waters. Nijhoff/Junk Publ. The Hague, pp. 448

Kramer, K.J.M., R.M. Warwick, & U.H. Brockmann, 1992. Manual on sampling and analytical procedures of tidal estuaries. JEEP92 report, IMW-TNO, Delft, pp. 238

Kramer, K.J.M. (ed), 1994. Biomonitoring of coastal waters and estuaries. CRC Press, Boca Raton, pp. 360

Laane, R.W., 1981. Composition and distribution of dissolved fluorescent substances in the Ems-Dollard estuary. Neth. J. Sea Res. 15: 88-99

Laane, R.W.P.M. & K.J.M. Kramer, 1990. Natural fluorescence in the North Sea and its major estuaries. Neth. J. Sea Res. 26: 1-9

Lauff, G.H. (ed), 1967. Estuaries. Am. Ass. Adv. Sci., Washington, DC, pp. 757

Laxen, D.P.H. & R.M. Harrison, 1981. Cleaning methods for polythene containers prior to the determination of trace metals in freshwater samples. Anal. Chem. 53: 345-350

Laxen, D.P.H. & I.M. Chandler, 1982. Comparison of filtration techniques for size distribution in fresh waters. Anal. Chem. 54: 1350-1355

Lazar, R. Edwards, R.C. Metcalfe, C.D. Metcalfe, T. Gobas, F.A.P.C. Haffner, G.D., 1992. A simple, novel method for the quantitative analysis of coplanar (nonorthosubstituted) polychlorinated biphenyls in environmental samples. Chemosphere, 25: 493-504

Leatherland, T.M., 1985. Operations in the field. In: Practical estuarine chemistry: a handbook. P.C. Head (ed). Cambridge Univ. Press, Cambridge, pp. 61-93

LeB. Williams, P.J., 1985. Analysis: organic matter. In: Practical estuarine chemistry: a handbook. P.C. Head (ed). Cambridge Univ. Press, Cambridge, pp. 160-200

Lee, C. & C. Cronin, 1982. The vertical flux of particulate organic nitrogen in the sea: decomposition of amino acids in the Peru upwelling area and the equatorial Atlantic. J. Mar. Res. 40: 227-251

Lee, C. & C. Cronin, 1984. Particulate amino acids in the sea: effects of primary productivity and biological decomposition. J. Mar. Res. 42: 1075-1097

Lee, M.L., M.V. Novotny & K.D. Bartle, 1981. Analytical chemistry of polycyclic aromatic compounds. Academic Press, New York

Leeder, M.R., 1982. Sedimentology. Process and product. Allen and Unwin, London, pp. 344

Legovic, T., 1991. Exchange of water in a stratified estuary with an application to Krka (Adriatic Sea). Mar. Chem. 32: 121-135

Legovic T., Z. Grzetic & A. Smircic, 1991. Effects of wind on a stratified estuary. Mar. Chem. 32: 153-161

Liebezeit, G., 1988. Distribution of dissolved humic compounds in the southern North Sea. In: Biogeochemistry and distribution of suspended matter in the North Sea and implications to fisheries biologyMitt. Geol. Palaeont. Inst. Univ. Hamburg, 85: 153-162

Lorenzen, C.J., 1967. Determination of chlorophyll and pheopigments: spectrophotometric equations. Limnol. Oceanogr. 12: 343-346

Loring, D.H., 1988. Normalisation of trace metal data. Report of the ICES working group on marine sediments in relation to pollution. ICES report C.M.1988/E:2 annex 3, Copenhagen

Loring, D.H. & R.T.T. Rantala, 1990. Sediments and suspended particulate matter: total and partial methods of digestion. Techniques in marine environmental sciences No. 9, ICES Copenhagen, pp. 14

Loring, D.H., 1991. Normalization of heavy-metal data from estuarine and coastal sediments. ICES J. mar. Sci. 48: 101-115

MacKinnon, M.D., 1981. The measurement of organic carbon in seawater. In: Marine organic chemistry, E.K. Duursma & R. Dawson (eds). Elsevier, Amsterdam, pp. 415-443

Mamayev, O.I., 1975. Temperature-salinity analysis of world ocean waters. Elsevier, Amsterdam, pp. 374

Mantoura, R.F.C. & C.A. Llewellyn, 1983. The rapid determination of algal chlorophyll and carotenoid pigments and their breakdown products in natural waters by reverse-phase high-performance liquid chromatography. Anal. Chim. Acta, 151: 297-314

Marshall, N., D.M. Skauwen, H.C. Lampe & C.A. Oviatt, 1973. Primary production of benthic microflora. In: A guide to the measurement of marine primary production under some special conditions. Unesco Monographs on Oceanographic Methodology, Unesco, Paris, 3: 37-44

Martin, D.F., 1968. Marine chemistry, Vol 1: Analytical methods. Marcel Dekker, New York, pp. 280

Mauchline, J., 1980. The biology of mysids and euphausiids. Adv. Mar. Biol. 18: 1-681

McDonald, P.D.M. & T.J. Pitcher, 1979. Age-groups from size-frequency data: a versatile and efficient method of analysing distribution mixtures. J. Fish. Res. Bd. Can. 36: 987-1001

McIntyre, A.D., J.M. Elliot & D.V. Ellis, 1984. Introduction: design of sampling programmes. In: Methods for the study of marine benthos. N.A. Holme & A.D. McIntyre (eds). 2nd ed. Blackwell, Oxford, pp. 1-26

McIntyre, A.D. & R.M. Warwick, 1984. Meiofauna techniques. In: Methods for the study of marine benthos. N.A. Holme & A.D. McIntyre (eds). 2nd ed. Blackwell, Oxford, pp. 217-244

Mees, J. & O. Hamerlynck, 1992. Spatial community of the winter hyperbenthos of the Schelde estuary, The Netherlands, and the adjacent coastal waters. Neth. J. Sea Res. 29: 357-370

Menzel, D.W. & R.F. Vaccaro, 1964. The measurement of dissolved organic and particulate carbon in sea water. Limnol. Oceanogr. 9: 138-142

Merian, E. (ed), 1991. Metals and their compounds in the environment. VCH Publ., Weinheim

 Millero, F.J., 1984. The conductivity-density-salinity-chlorinity relationships for estuarine waters. Limnol. Oceanogr. 29: 1317-1321

Mopper, K., 1977. Sugars and uronic acids in sediment and water from the Black Sea and North Sea with emphasis on analytical techniques. Mar. Chem. 5: 585-603

Mopper, K., 1978. Improved chromatographic separation on anion exchange resins. Anal. Biochem. 87: 162-168

Mopper, K., R. Dawson, G. Liebezeit & H.P. Hansen, 1980. Borate complex ion exchange chromatography with fluorometric detection for determination of saccharides. Anal. Chem. 52: 2018-2022

Morin, A., 1985. Variability of density estimates and the optimazation of sampling programs for stream benthos. Can. J. Fish. aquat. Sci. 42: 1530-1534

Morris, A.W. (ed), 1983. Practical procedures for estuarine studies. A handbook prepared by the Estuarine Ecology Group of the Institute for Marine Environmental Research. IMER, Plymouth, pp. 262

Morris, A.W., 1983. Strategy for practical estuarine studies. In: Practical procedures for estuarine studies, A.W. Morris (ed), Institute for Marine Environmental Research, Plymouth, pp. 1-17

Morris, A.W., 1985. Estuarine chemistry and general survey strategy. In: Practical estuarine chemistry: a handbook. P.C. Head (ed). Cambridge Univ. Press, Cambridge, pp. 1-60

Morris, A.W., 1990. Guidelines for monitoring estuarine waters and suspended matter. UNEP Regional Seas Programme. Reference methods for Marine Pollution Studies. In press

Morris, A.W. & J.P. Riley, 1966. The bromide/chlorinity and sulphate/chlorinity ratio in sea water. Deep Sea Res. 13: 699-705

Mudroch, A. & R.A. Bourbonniere, 1991. Sediment sample handling and processing. In: Handbook of techniques for aquatic sediments sampling; A. Mudroch & S.D. MacKnight (eds). CRC Press, Boca Raton, pp. 131-169

Mudroch, A. & S. MacKnight, 1991. Bottom sediment sampling. In: Handbook of techniques for aquatic sediments sampling; A. Mudroch & S.D. MacKnight (eds). CRC Press, Boca Raton, pp. 29-96

Müller, G., 1964. Sediment Petrologie. I. Methoden der Sedimentationsanalyse. Scheizerbart, Stuttgart

Murray, A.P., C.F. Gibbs, A.R. Longmore & D.J. Flett, 1986. Determination of chlorophyll in marine waters: intercomparison of a rapid HPLC method with full HPLC, spectrophotometric and fluorometric methods. Mar. Chem. 19: 211-227

Murray, A.P., B.J. Richardson & C.F. Gibbs, 1991. Bioconcentration factors for petroleum hydrocarbons, PAHs, LABs and biogenic hydrocarbons in the blue mussel. Mar. Pollut. Bull. 22: 595-603

Neff, J.M., 1979. Polycyclic aromatic hydrocarbons in the aquatic environment: Sources, fates and biological effects. Applied Science Publishers

Neilson, B.J., A. Kuo & J. Brubaker (eds), 1989. Estuarine circulation. Humana Press, Clifton, NJ, pp. 400

Nisbet, I.C.T. & P. LaGoy, 1992. Toxic Equivalent Factors (TEFs) for polycyclic aromatic hydrocarbons (PAHs). Regul. Toxicol. Pharmacol. 16: 290-300

Nusch, E.A., 1980. Comparison of different methods for chlorophyll and phaeopigment determination. Arch. Hydrobiol. Beih. Ergebn. Limnol. 14: 14-36

O'Connor, T.P., A.Y. Cantillo & G.G. Lauenstein, 1994. Monitoring of temporal trends in chemical contamination by NOAA National Status and Trends Mussel Watch Project. In: Biomonitoring of coastal waters and estuaries, K.J.M. Kramer (ed). CRC Press, Boca Raton, pp. 29-50

Olausson, E. & I. Cato (eds), 1980. Chemistry and biogeochemistry of estuaries. Wiley, Chichester, pp. 452

Omori, M. & T. Ikeda, 1984. Methods in marine zooplankton ecology. Wiley, New York, pp. 322

Parsons, T.R., Y. Maita & C.M. Lalli, 1984. A manual of chemical and biological methods for seawater analysis. Pergamon, Oxford, pp. 173

Perkins, E.J., 1974. The biology of estuaries and coastal waters. Acad. Press, London

Pfannkuche, O. & H. Thiel, 1988. Sample processing. In: Introduction to the study of meiofauna. R.P. Higgins & H. Thiel (eds). Smithsonian Inst. Press, Washington DC, pp. 134-145

Phillips, D.J.H., 1994. Macrophytes as biomonitors of trace metals. In: Biomonitoring of coastal waters and estuaries, K.J.M. Kramer (ed). CRC Press, Boca Raton, pp. 85-106

Pierce, J.B. & L. Despres-Patanjo, 1988. A review of monitoring strategies and assessments of estuarine pollution. Aquat. Toxicol. 11: 323-343,

Pringle, J.D., 1984. Efficiency estimates for various quadrat sizes used in benthic sampling. Can. J. Fish. Aquat. Sci. 41: 1485-1489

Pritchard, D.W., 1955. Estuarine circulation patterns. Proc. Am. Soc. civ. Engrs. 81: 1-11

Pritchard, D.W., 1967. Observations of circulation in coastal plain estuaries. In: Estuaries, G.H. Lauff (ed). Am. Ass. Adv. Sci., Washington, DC, pp. 37-44

Raymont, J.E.G., 1983. Plankton and productivity in the oceans. 2nd ed. Volume 2. Zooplankton. Pergamon press, Oxford, pp. 824

Reemtsma, T., B. Haake, V. Ittekot, U.H. Brockmann & R.R. Nair, 1990. Downward flux of particulate fatty acids in the central Arabian Sea. Mar. Chem. 29: 183-202

Reuter, R., 1980. Characterization of marine particle suspensions by light scattering. I. Numerical predictions from Mie theory II. Experimental results. Oceanologica Acta 3: 317-324; 325-332

Reutergård, L., 1980. Chlorinated hydrocarbons in estuaries. E. Olausson & I. Cato (eds.). Chemistry and biogeochemistry of estuaries. John Wiley & Sons, Chichester, pp. 349-366

Riddle, M.J., 1989a. Precision of the mean and the design of benthic sampling programmes: caution advised. Mar. Biol. 103: 225-230

Riddle, M.J., 1989b. Bite profiles of some benthic grab samplers. Est. Coast. Shelf Sci. 29: 285-292

Riley, G.A., 1970. Particulate organic matter in sea water. Adv. Mar. Biol. 8: 1-118

Robinson, G.A. & A.R. Hiby, 1978. Sampling techniques: the continuous Plankton Recorder survey. In: Phytoplankton manual. A. Sournia (ed). Unesco, Paris, pp. 59-63

Rodier, J. (ed), 1984. L'analyse de l'eau: eaux naturelles, eaux résiduaires, eau de mer. 7th Ed., Dunod, Paris, pp. 1365

Round, F.E. & M. Hickman, 1984. Phytobenthos sampling and estimation of primary production. In: Methods for the study of marine benthos. N.A. Holme & A.D. McIntyre (eds). 2nd ed. Blackwell, Oxford, pp. 245-283

Safe, S., 1990. Polychlorinated biphenyls (PCBs), dibenzo-p-dioxins (PCDDs), dibenzofurans (PCDFs), and related compounds: Environmental and mechanistic considerations which support the development of toxic equivalency factors (TEFs). Crit. Rev. Toxicol. 21: 51-88

Salim, R. & B.G. Cooksey, 1981. The effect of centrifugation on the suspended particles of river waters. Water Res. 15: 835-839.

Saliot, A., J. Trouczynski, P. Scribe & R. Letolle, 1988. The application of isotopic and biogeochemical markers to the study of the biogeochemistry of organic matter in a macrotidal estuary, the Loire. Est. Shelf Sci. 27: 645-660

Saliot, A., 1994. Cours de biogéochimie organique marine. Océanis, Inst. Océanographique, Paris

Salomons, W. & U. Förstner, 1984. Metals in the hydrocycle. Springer, Berlin, pp. 349.

Salonen, K. & J. Sarvala, 1985. Combination of freezing and aldehyde fixation. A superior preservation method for biomass determination of aquatic invertebrates. Arch. Hydrobiol. 10: 217-230

Sfriso, A., S. Raccanelli, B. Pavoni & A. Marcomini, 1991. Sampling strategies for measuring macroalgal biomass in the shallow waters of the Venice lagoon. Environm. Technol. 12: 263-269

Sharp, J.H., 1974. Improved analysis for particulate organic carbon and nitrogen from sea water. Limnol. Oceanogr. 19: 984-989

Sournia, A. (ed), 1978. Phytoplankton manual. Unesco, Paris, pp. 337

Spencer, M.J., S.R. Piotrowicz & P.R. Betzer, 1982. Concentrations of cadmium, copper, lead and zinc in surface waters of the northwest Atlantic Ocean - a comparison of Go-flo and teflon water samplers. Mar. Chem. 11: 403-410.

Stoeppler, M., 1991. Analytical chemistry of metals and metal compounds. Metals and their compounds in the environment, E. Merian (ed.). VCH, Weinheim, pp. 105-206

Rumohr, H., 1990. Soft bottom macrofauna: collection and treatment of samples. Techniques in marine environmental sciences, 8. ICES, Copenhagen, pp. 18

Say, P.J., I.G. Burrows & B.A. Whitton, 1986. *Enteromorpha* as a monitor of heavy metals in estuarine and coastal intertidal waters. A method for the sampling, treatment and analysis of the seaweed *Enteromorpha* to monitor heavy metals in estuaries and coastal waters. Occasional publication No. 1, Northern Environmental Consultants Ltd., Consett, Co. Durham, pp. 25

Sorbe, J.C., 1983. Description d'un traîneau destiné à l'échantillonage quantitatif étagé de la faune suprabenthique néritique. Ann. Inst. Océanogr. 59: 117-126

Steedman, J.H. (ed), 1976. Zooplankton fixation and preservation. Unesco monogr. oceanogr. methodol. 4. Unesco, Paris, pp. 350

Strathmann, R.R., 1967. Estimating the organic carbon content of phytoplankton from cell volume or plasma volume. Limnol. Oceanogr. 12: 411-418

Strickland, J.D.H. & T.R. Parsons, [1968], 1972. 2nd edition. A practical handbook of seawater analysis. Fish. Res. Bd. Can. Bulletin 167, Ottawa, pp. 311

Sugimura Y. & Y. Suzuki, 1988. A high-temperature catalytic oxidation method for the determination of non-volatile dissolved organic carbon in seawater by direct injection of a liquid sample. Mar. Chem. 24: 105-131

Suzuki, Y. & Y. Sugimura, 1985. A catalytic oxidation method for the determination of total nitrogen dissolved in seawater. Mar. Chem. 16: 83-97

Suzuki, Y. Tanoue, E. Ito, H., 1992. A high-temperature catalytic oxidation method for the determination of dissolved organic carbon in seawater - analysis and improvement. Deep Sea Res. A 39: 185-198

Tabatabai, M.A., 1974. Determination of sulphate in water samples. Sulphur Inst. J. 10: 11-13

Tangen, K., 1978. Sampling techniques: nets. In: Phytoplankton manual. A. Sournia (ed). Unesco, Paris, pp. 50-58

Tett, P.B., 1987. Plankton. In: Biological surveys of estuaries and coasts, J.M. Baker, & W.J. Wolff (eds). EBSA Handbook, Cambridge Univ. Press, Cambridge, pp. 280-341

Thistle, D. & J.W. Fleeger, 1988. Sampling strategies. In: Introduction to the study of meiofauna, R.P. Higgins & H. Thiel (eds). Smithsonian Inst. Press, Washington DC, pp. 126-133

Throndsen, J., 1978. Preservation and storage. In: Phytoplankton manual. A. Sournia (ed). Unesco, Paris, pp. 69-74

Toggweiler, J.R., 1988. Deep sea carbon, a burning issue? Nature, 334: 468

Toggweiler, J.R., 1989. Is the downward dissolved organic matter (DOM) flux important in carbon transport? In: Productivity of the ocean: Present and Past, W.H. Berger, V.S. Smetacek & G. Wefer (eds.). Wiley, New York, pp. 65-83

Tranter, D.J. & J.H. Fraser (eds), 1968 (1974 2nd ed). Zooplankton sampling. Unesco Monographs on oceanographic methodology, 2. Unesco, Paris, pp. 174

Uhlig, G.H., H. Thiel & J.S. Gray, 1973. The quantitative separation of meiofauna. A comparison of methods. Helgol. wiss. Meeresunt. 25: 173-195

Uncles, R.J., M.B. Jordan & R.C.A. Elliot, 1983. Sampling and analysis of physical features. In: Practical procedures for estuarine studies, A.W. Morris (ed), Institute for Marine Environmental Research, Plymouth, pp. 19-53

UNEP/FAO/IAEA, 1984. Sampling of selected marine organisms and sample preparation for the analysis of chlorinated hydrocarbons. In: UNEP/FAO/ IAEA/ IOC Reference methods for marine pollution studies No. 12, Rev. 1, UNEP, Nairobi

UNEP/IOC/IAEA, 1990. Determination of dissolved/dispersed hydrocarbons in marine waters. In: UNEP report, reference methods for marine pollution studies No. 56, UNEP, Nairobi

Unesco, 1981. The practical salinity scale 1978 and the International Equation of state of seawater 1980. Technical Papers in Marine Science, no. 36, pp. 25

Unesco, 1985. The International System of Units (SI) in oceanography. Technical Papers in Marine Science, no. 45, pp. 124

Vale, C. & B. Sundby, 1987. Suspended sediment fluctuations in the Tagus estuary on semi-diurnal and fortnightly time scales. Est. Coast. Shelf Sci. 25: 495-508

Van den Berg, C.M.G., S.H. Khan, P.J. Daly, J.P. Riley & D.R. Turner, 1991. An electrochemical study of Ni, Sb, Se, Sn, U and V in the estuary of the Tamar. Est. Coast. Shelf Sci. 33: 309-322

Van den Hoek, C., W. Admiraal, F. Colijn &. V.N. de Jonge, 1979. The role of algae and seagrasses in the ecosystem of the Wadden Sea: a review. In: Ecology of the Wadden Sea, Vol. 1, W.J. Wolff (ed), Balkema, Rotterdam, pp. 3/9-3/118

Van der Veer, J., 1982. Simple and reliable methods for the fixation, mounting and staining of small and delicate marine plankton for light microscopic identification. Mar. Biol. 66: 9-14

Venrick, E.L., 1978. Sampling techniques: water bottles. In: Phytoplankton manual. A. Sournia (ed). Unesco, Paris, pp. 33-40

Venrick, E.L., 1978. Sampling strategies. In: Phytoplankton manual. A. Sournia (ed). Unesco, Paris, pp. 7-16

Vézina, A.F., 1988. Sampling variance and the design of quantitative surveys of the marine benthos. Mar. Biol. 97: 151-155

Waid, J.S., 1986-1987. PCBs and the environment (3 Vols). CRC Press, Boca Raton Fl.

Vollenweider, R.A., 1969. A manual on methods for measuring primary production in aquatic environments. IBP Handbook, Blackwell, Oxford, pp. 213

Warwick, R.M., 1984. Sampling and analysis of benthic communities. In: Practical procedures for estuarine studies, A.W. Morris (ed). PML, Plymouth, pp. 185-212

Wentworth, C.K., 1922. A scale of grade and class terms for clastic sediments. J. Geol. 30: 377-392

Widbom, B., 1984. Determination of average individual dry-weights and ash-free dry-weights in different sieve fractions of marine meiofauna. Mar. Biol. 84: 101-108

Wiltshire, K.H., 1991. Experimental procedures for the fractionation of phosphorus in sediments with emphasis on anaerobic techniques. Verh. Internat. Verein. Limnol. 24: 3073-3078

Wolff, W.J., 1973. The estuary as a habitat. An analysis of data on the soft-bottom macrofauna of the estuarine area of the rivers Rhine, Meuse, and Scheldt. Zool. Verhand. no. 126, pp. 242

Wolff, W.J. (ed), 1983. Flora and vegetation of the Wadden Sea. In: Ecology of the Wadden Sea, Vol. 1, W.J. Wolff (ed), Balkema, Rotterdam, pp. 3/1-3/206

Wolff, W.J., 1987a. Flora and fauna of intertidal sediments. In: Biological surveys of estuaries and coasts, J.M. Baker, & W.J. Wolff (eds). EBSA Handbook, Cambridge Univ. Press, Cambridge, pp. 81-105

Wolff, W.J., 1987b. Identification. In: Biological surveys of estuaries and coasts, J.M. Baker, & W.J. Wolff (eds). EBSA Handbook, Cambridge Univ. Press, Cambridge, pp. 404-423

Wollast, R. & J.C. Duinker, 1982. General methodology and sampling strategy for studies on the behaviour of chemicals in estuaries. Thalassia Jugosl. 18: 471-491

Wong, C.W., K. Kremling, J.P. Riley, W.K. Johnson, V. Stukas, P.G. Berrang, P. Erickson, D. Thomas, H. Petersen & B. Imber, 1983. An intercomparison of sampling devices and analytical techniques using seawater from a CEPEX enclosure. In: Trace metals in seawater; C.S. Wong, E. Boyle, K.W. Bruland, J.D. Burton & E.D. Goldberg (eds.). Plenum, New York

Yamada, H. & M. Kayama, 1987. Distribution and dissolution of several forms of phosphorus in coastal marine sediments. Oceanol. Acta, 10: 311-321

Yeats, P.A., 1987. Trace metals in sea water; sampling and storage methods. ICES Techniques in marine environmental sciences No. 2, Copenhagen, pp. 8.

Yeats, P.A. & L. Brügmann, 1990. Suspended particulate matter: collection methods for gravimetric and trace metal analysis. ICES Techniques in marine environmental sciences No. 7, Copenhagen, pp. 9

Zimmerman, J.T.F. & J.W. Rommets, 1974. Natural fluorescence as a tracer in the Dutch Wadden Sea. Neth. J. Sea Res. 8: 117-125

Zutic, V. & T. Legovic, 1987. A film of organic matter at the freshwater/seawater interface of an estuary. Mar. Chem. 32: 163-170

Zutic, V. (ed), 1991. Physical, chemical and biological processes in stratified estuaries. Mar. Chem. 32(2-4), special volume, pp. 111-390

# ANNEXES

# Annex I

## Forms for the inventory of European tidal estuaries

(De Winter, 1992b)

# Synopsis of the Ecology and Hydrography of European Estuaries

Name of the Estuary ........................................................................................................

Geographical Coordinates ........................................................................................................

Countries ........................................................................................................

Type of the Estuary    [   ]

River(s) discharging ........................................................................................................

Human interference (like dredging): ........................................................................................................

General description ........................................................................................................

........................................................................................................

........................................................................................................

........................................................................................................

........................................................................................................

........................................................................................................

**Additional Maps:**    O   Estuary    O   Sediment    O   Catchment Area   O   Other

Maps available at ........................................................................................................

under number ........................................................................................................

| **Form completed by:** | |
|---|---|
| **Institute:** | .......................................................................... |
| **adress:** | .......................................................................... |
| | .......................................................................... |
| | .......................................................................... |
| | .......................................................................... |
| **telephone nr.:** | .......................................................................... |
| **fax nr.:** | .......................................................................... |

## Economic Use:

| | |
|---|---|
| Industrial Zones | .................................................................................................................... |
| Big cities | .................................................................................................................... |
| Fisheries | .................................................................................................................... |
| Aqua-culture | .................................................................................................................... |

## General Physical Characteristics:

| | mean | unit | minimum | maximum | also enclosed: time series | graph |
|---|---|---|---|---|---|---|
| Catchment area | ............... | km² | | | | |
| Water volume | ............... | * $10^6$ m³ | | | | |
| Residence time | ............... | days | ............... to | ............... | | |
| Number of ice days | ............... | | ............... to | ............... | O | O |
| Water temperature (seaward part) | ............... | °C | ............... to | ............... | O | O |
| Position of turbidity maximum (from mouth) | ............... | km | ............... to | ............... | O | O |
| Mean tidal ranges | ............... | m at | ............... | | | |
| | ............... | m at | ............... | | | |
| | ............... | m at | ............... | | | |
| | ............... | m at | ............... | | | |

## Hydrography

| | Freshwater | Oligohaline | Mesohaline | Polyhaline | Euhaline | Whole Estuary | |
|---|---|---|---|---|---|---|---|
| Water surface at MHW: | ............... | ............... | ............... | ............... | ............... | ............... | km² |
| Length of section at MHW: | ............... | ............... | ............... | ............... | ............... | ............... | km |
| Width of section at MHW: | ............... | ............... | ............... | ............... | ............... | ............... | km |
| Salt marshes at MHW: | ............... | ............... | ............... | ............... | ............... | ............... | km² |
| 0 - 5 m depth at MLW: | ............... | ............... | ............... | ............... | ............... | ............... | km² |
| 5 -10 m depth at MLW: | ............... | ............... | ............... | ............... | ............... | ............... | km² |
| >10 m depth at MLW: | ............... | ............... | ............... | ............... | ............... | ............... | km² |
| Water volume: | ............... | ............... | ............... | ............... | ............... | ............... | $10^6$m³ |
| Residual current: | ............... | ............... | ............... | ............... | ............... | ............... | m/s |

# Hydrography

| | Freshwater | Oligohaline | Mesohaline | Polyhaline | Euhaline | Whole Estuary |
|---|---|---|---|---|---|---|
| **areas of intertidal flats (all at MLW):** | | | | | | |
| — total surface: | .................. | .................. | .................. | .................. | .................. | .................. km² |
| — Sandy flats (<5% clay) | .................. | .................. | .................. | .................. | .................. | .................. km² |
|   < 50% of time emerged | .................. | .................. | .................. | .................. | .................. | .................. km² |
|   ≥ 50% of time emerged | .................. | .................. | .................. | .................. | .................. | .................. km² |
| — Silty (>5% clay) | .................. | .................. | .................. | .................. | .................. | .................. km² |
|   < 50% of time emerged | .................. | .................. | .................. | .................. | .................. | .................. km² |
|   ≥ 50% of time emerged | .................. | .................. | .................. | .................. | .................. | .................. km² |
| — Beds of bivalves: | .................. | .................. | .................. | .................. | .................. | .................. km² |
| — Beds of vegetation: | .................. | .................. | .................. | .................. | .................. | .................. km² |

# Biology

| Eco-groups: | Freshwater | Oligohaline | Mesohaline | Polyhaline | Euhaline | not specified |
|---|---|---|---|---|---|---|
| Phytoplanktn Biom.: | .......... | .......... | .......... | .......... | .......... | .......... gC/m³ |
|   Production: | .......... | .......... | .......... | .......... | .......... | .......... gC/m³.y |
|   remarks: | | | | | | |
| Zooplankton Biom.: | .......... | .......... | .......... | .......... | .......... | .......... gC/m³ |
|   Production: | .......... | .......... | .......... | .......... | .......... | .......... gC/m³.y |
|   remarks: | | | | | | |
| Phytobenthos Biom.: | .......... | .......... | .......... | .......... | .......... | .......... gC/m² |
|   Production: | .......... | .......... | .......... | .......... | .......... | .......... gC/m².y |
|   remarks: | | | | | | |
| Macrobenths Biom.: | .......... | .......... | .......... | .......... | .......... | .......... gC/m² |
|   Production: | .......... | .......... | .......... | .......... | .......... | .......... gC/m².y |
|   remarks: | | | | | | |
| Meiobenthos Biom.: | .......... | .......... | .......... | .......... | .......... | .......... gC/m² |
|   Production: | .......... | .......... | .......... | .......... | .......... | .......... gC/m² y |
|   remarks: | | | | | | |
| Hyperbenthos Biom.: | .......... | .......... | .......... | .......... | .......... | .......... gC/m² |
|   Production: | .......... | .......... | .......... | .......... | .......... | .......... gC/m².y |
|   remarks: | | | | | | |

**Biological References:** [1]

| | species distribution/composition | | energetics | | |
|---|---|---|---|---|---|
| | spatial | temporal | biomass | production | consumption |
| phytoplankton | | | | | |
| zooplankton | | | | | |
| phytobenthos | | | | | |
| meiobenthos | | | | | |
| macrozoobenthos | | | | | |
| hyperbenthos | | | | | |
| Birdlife | | | | | |
| Wildlife | | | | | |

[1] Please refer to numbers in the reference list you have included.


## Physico-Chemical Characteristics:

| | time series enclosed | graph of series enclosed [1] | reference(s) |
|---|---|---|---|
| Oxygen | ○ | ○ | |
| Salinity/Chlorinity | ○ | ○ | |
| Nitrogen | ○ | ○ | |
| Phosphorus | ○ | ○ | |
| Silicate | ○ | ○ | |
| DOC | ○ | ○ | |
| POC | ○ | ○ | |
| organic load in IE | ○ | ○ | |
| light intensity $I_0$ | ○ | ○ | |
| light extinction | ○ | ○ | |
| temperature | ○ | ○ | |
| | ○ | ○ | |
| | ○ | ○ | |

# Annex    II

# Synopsis of sampling methods and sampling (sub)codes

## Physico-chemical variables water (S-6.1)

| Code | Method | subcode sample: | |
|------|--------|---------|---|
| 1a | water sampler | 1 | spot |
| 1b | sample bottle | 2 | integrated |
| 2 | pumping system | 3 | profile |
| 3 | bucket | 0 | other |
| 0 | other | | |

## Physico-chemical variables seston (S-6.2)

| Code | Method | subcode sample: | | subcode filter size: | |
|------|--------|---------|---|---------|---|
| 1a | water sampler | 1 | spot | 1a | 0.45 µm |
| 1b | sample bottle | 2 | integrated | 1b | 0.4 µm |
| 2 | pumping system | 3 | profile | 2 | centrifugation |
| 3 | bucket | 0 | other | 0 | other |
| 0 | other | | | | |

## Physico-chemical variables sediment  (S-6.3)

| Code | Method | subcode sample: | |
|------|--------|---------|---|
| 1a | handcorer | 1 | mixed |
| 1b | gravity corer | 2 | profile |
| 1c | piston corer | 0 | other |
| 1d | box corer | | |
| 1e | vibro corer | | |
| 2 | grab | | |
| 0 | other | | |

---

## Biota: water

**Code  Method**                    **subcode**                    **subcode**

### bacteria (S-6.4)

| 1 | sample bottle |
|---|---------------|
| 2 | pumping system |
| 3 | (sterile) water sampler |
| 0 | other |

### phytoplankton (S-6.5)

| 1a | sample bottle |
|----|---------------|
| 1b | water sampler |
| 2 | pumping system |
| 3 | nets |
| 0 | other |

### zooplankton (S-6.6)

| Code | Method | net-size: | | volume: | |
|------|--------|-----------|---|---------|---|
| 1 | pump + net | 1 | 55 or 63 μm | 1 | 100 l |
| 2 | towing net | 2 | 200 μm | 2 | 5 m³ |
| 3 | high-speed sampler | 0 | other | 0 | other |
| 4 | tube + net | | | | |
| 5 | bucket + net | | | | |
| 0 | other | | | | |

| Code | Method | subcode | | subcode |
|------|--------|---------|--|---------|

### hyperbenthos (S-6.7)

|   |   |   | mesh-size: |
|---|---|---|------------|
| 1 | sledge | 1 | 1 mm |
| 2 | passive fishing | 2 | 0.5 mm |
| 3 | high-speed sampler | 0 | other |
| 0 | other | | |

---

# Biota: sediment

| Code | Method | subcode | | subcode |
|------|--------|---------|--|---------|

### micro-phytobenthos (S-6.8)

| 1 | hand corer 1 cm ø (1 cm depth) |
|---|--------------------------------|
| 2 | box corer, subsamples |
| 0 | other |

### macro-phytobenthos (S-6.9)

|   |   |   | area: |
|---|---|---|-------|
| 1 | grid-squares | 1 | 10 * 10 cm |
| 0 | other | 2 | 25 * 25 cm |
|   |   | 0 | other |

### meiobenthos (sub-tidal) (S-6.10)

|    |    |   | depth: |   | sieve mesh size: |
|----|----|---|--------|---|------------------|
| 1a | box corer (2 cm Ø, mud) | 1 | 5 cm | 1 | 63 µm |
| 1b | box corer (8 cm Ø, sand ) | 2 | 15 cm | 2 | 45 µm |
| 2a | divers (2 cm Ø, mud) | 0 | other | 0 | other |
| 2b | divers (8 cm Ø, samd) | | | | |
| 3  | bow wave free sampler | | | | |
| 4  | pole, gravity, piston corers | | | | |
| 5  | grab sampler | | | | |
| 0  | other | | | | |

### meiobenthos (inter-tidal) (S-6.10)

|    |    |   | depth: |   | sieve mesh size: |
|----|----|---|--------|---|------------------|
| 1a | hand corer (2 cm ø, mud) | 1 | 5 cm | 1 | 63 µm |
| 1b | hand corer (8 cm ø, sand) | 2 | 15 cm | 2 | 45 µm |
| 0  | other | 0 | other | 0 | other |

| Code | Method | | subcode | | | subcode |
|------|--------|---|---------|---|---|---------|

**macro-zoobenthos (sub-tidal) (S-6.11)**

| | | | **depth:** | | | **sieve mesh size:** |
|---|---|---|---|---|---|---|
| 1 | box corer | 1 | 25 cm | 1 | 1 mm |
| 2 | grab sampler | 0 | other | 2 | 0.5 mm |
| 0 | other | | | 0 | other |

**macro-zoobenthos (inter-tidal) (S-6.11)**

| | | | **depth:** | | | **sieve mesh size:** |
|---|---|---|---|---|---|---|
| 1 | hand corer (200 cm$^2$) | 1 | 25 cm | 1 | 1 mm |
| 2 | hand corer (other size) | 0 | other | 2 | 0.5 mm |
| 0 | other | | | 0 | other |

# *In situ* measurements: water

**Code  Method**                    **subcode**

### salinity

1    conductivity
2    lab. analysis
0    other

### temperature

1    thermocouple
2    thermometer
0    other

### pH

1a   *in situ* pH-electrode
1b   on-board determination
2    lab. analysis
0    other

### oxygen

1    *in situ* oxygen electrode
2    lab. analysis
0    other

### light penetration

1    Secchi disc
2    irradiance meter
0    other

### turbidity

|   |   | **cell:** |   |
|---|---|---|---|
| 1  | *in situ* OBS | 1 | 5 cm |
| 2  | *in situ* transmissometer | 2 | 10 cm |
| 3a | *in situ* nephelometer | 0 | other |
| 3b | pumping system/ on board analysis |   |   |
| 0  | other |   |   |

### fluorescence

|   |   | **cell:** |   |
|---|---|---|---|
| 1 | *in situ* fluorimeter | 1 | 5 cm |
| 2 | pumping system/ on board analysis | 2 | 1 cm |
|   |   | 0 | other |
| 0 | other |   |   |

# Annex III

## Synopsis of analytical methods and methods (sub)codes

**Physico-chemical variables: water**

**Code Method**　　　　　　**subcode**

*salinity  (A-7.1)*

1a  conductivity
1b  chlorinity
0   other

*chlorinity  (A-7.2)*

1   titration
0   other

*temperature (A-7.3)*

1a  thermistor
1b  resistance thermometer
2   thermometer
0   other

*turbidity (A-7.4)*

　　　　　　　　　　　　　　**cell length:**
1   Secchi depth　　　　1　　1 cm
2   beam transmission　　2　　5 cm
3   nephelometer　　　　3　　10 cm
0   other　　　　　　　　0　　other

| Code | Method | subcode |
|------|--------|---------|

### dissolved oxygen (A-7.5)

| | |
|---|---|
| 1a | Winkler titration |
| 1b | oxygen electrode |
| 0 | other |

### pH (A-7.6)

| | |
|---|---|
| 1 | pH electrode |
| 0 | other |

### total alkalinity (A-7.7)

| | |
|---|---|
| 1 | pH method |
| 2 | Gripenberg method |
| 3 | potentiometric titration |
| 0 | other |

### nitrate (A-7.8)

**equipment:**

| | | | |
|---|---|---|---|
| 1 | reduction method | 1 | automated |
| 0 | other | 2 | manual |

### nitrite (A-7.9)

**equipment:**

| | | | |
|---|---|---|---|
| 1 | sulphanilamide method | 1 | automated |
| 0 | other | 2 | manual |

### ammonia (A-7.10)

**equipment:**

| | | | |
|---|---|---|---|
| 1 | oxidation method | 1 | automated |
| 2 | indophenol method | 2 | manual |
| 0 | other | | |

### phosphate (A-7.11)

**equipment:**

| | | | |
|---|---|---|---|
| 1 | molybdic acid method | 1 | automated |
| 0 | other | 2 | manual |

### silicate (A-7.12)

**equipment:**

| | | | |
|---|---|---|---|
| 1a | bleu colour method | 1 | automated |
| 1b | yellow colour method | 2 | manual |
| 0 | other | | |

## Code  Method

### *hydrogen sulphide (A-7.13)*

1a   methylene blue method
1b   Lauth's violet method
2    manganese method
3    ion specific electrode
0    other

### *sulphate (A-7.14)*

1    turbidimetric method
2    gravimetric method
3    methylthymol method
4    sulphide method
0    other

### *dissolved organic carbon (A-7.15)*

1    high temp. catalytic oxidation (Pt, $CO_2$)
2    high temp. catalytic oxidation (Cu, $CH_4$)
3    high temp. catalytic oxidation (Ni, $CH_4$)
4    wet chemical oxidation
5    UV oxidation
6    dry combustion
0    other

### *dissolved organic nitrogen (A-7.16)*

1    persulphate oxidation
2    micro Kjeldahl nitrogen
3    high temp. catalytic oxidation (Pt, $N_2O$)
4    UV oxidation
0    other

### *dissolved humic compounds (yellow acids) (A-7.17)*

1    fluorescence, filter
2    fluorescence scan
0    other

### *(dissolved) total carbohydrates (A-7.18)*

1    tryptophan method
0    other

**Code  Method**

*individual carbohydrates (A-7.19)*

1    HPLC
2    $C_{18}$ column
3    reduction method
0    other

*(dissolved) total amino acids (A-7.20)*

1    fluorimetric method
0    other

*individual amino acids and -sugars (A-7.21)*

1    liquid chromatography
2    HPLC
0    other

*(dissolved) proteins (A-7.22)*

1    brilliant blue method
2    fluorimetric method
0    other

*(dissolved) lipids (A-7.23)*

1    GC-MS
0    other

*(dissolved) trace metals (A-7.24)*

1a   complexation/extraction, (ET)AAS
1b   complexation/extraction, (ZET)AAS
1c   complexation/extraction, ICP-MS
1d   complexation/extraction, ICP-AES
2    anodic stripping voltammetry (DPASV)
3    cathodic stripping voltammetry (ADPCSV)
0    other

*(dissolved) PAHs (A-7.25)*

1a   RP-HPLC, without filtration, total
1b   RP-HPLC, with filtration, total
1c   RP-HPLC, after solid phase extraction (dissolved)
2a   GC-MS, without filtration, total
2b   GC-MS, with filtration, total
2c   GC-MS, after solid phase extraction (dissolved)
3a   ID-GC-MS, without filtration, total
3b   ID-GC-MS, with filtration, total
3c   ID-GC-MS, after solid phase extraction (dissolved)
0    other

**Code  Method**

*(dissolved) PCBs (A-7.26)*

1a   GC-ECD, total
1b   GC-ECD, dissolved
2a   GC-MS, total
2b   GC-MS, dissolved
3a   ID-GC-MS, total
3b   ID-GC-MS, dissolved
0    other

## Physico-chemical variables: seston

**Code  Method**                    **subcode**

### suspended particulate matter (A-7.27)

1    filtration
2    centrifugation
3    calibrated turbidity
0    other

### particle size (per size class) (A-7.28)

1    conductometric analysis
2    laser diffraction
3    flow-cytometry
0    other

### pigments (A-7.29)

|   |   |   | **extractant:** |
|---|---|---|---|
| 1 | inverse HPLC | 1 | ethanol 90% |
| 2 | spectrophotometry | 2 | acetone 90% |
| 3 | fluorimetric analysis | 3 | methanol 90% |
| 0 | other | 0 | other |

### particulate organic carbon (A-7.30)

1    dry combustion
2    persulphate oxidation
3    spectrophotometry
0    other

### particulate organic nitrogen (A-7.31)

1    dry combustion method
2    total Kjeldahl nitrogen
3    persulphate digestion
0    other

### particulate organic phosphorus (A-7.32)

1    persulphate digestion
2    sulphuric acid/peroxide digestion
3    perchloric acid digestion
0    other

# Physico-chemical variables: sediment

## Code  Method

### clay  content (A-7.33)

1a  pipette method, peroxide
1b  pipette method, no peroxide
2   Atterberg method
3   conductometric analysis
4   laser diffraction
0   other

### silt content (A-7.34)

1   wet sieving
2   conductometric analysis
3   laser diffraction
0   other

### grain size distribution (A-7.35)

1a  dry sieving
1b  wet sieving
0   other

### particulate trace metals (A-7.36)

1   total digestion, total sediment
2   total digestion, <63 $\mu$m fraction
3   partial digestion, total sediment
4   partial digestion, <63 $\mu$m fraction
0   other

### particulate PAHs (A-7.37)

1   RP-HPLC detection
2   GC-MS detection
3   ID-GC-MS detection
0   other

### particulate PCBs (A-7.38)

1   GC-ECD detection
2   GC-MS detection
3   ID-GC-MS detection
0   other

# Biological variables: water

## Code  Method

### bacteria

*total numbers (A-7.39)*
1    fluorescence microscopy
0    other

*production (A-7.40)*
1    thymidine method
2    growth rate
3    dialysis method
0    other

### phytoplankton

*species abundance (A-7.41)*
1a   microscopy, after concentration
1b   microscopy, without concentration
0    other

*production (A-7.42)*
1a   $^{14}C$ method, incubator
1b   $^{14}C$ method, *in situ*
2a   oxygen determination method
2b   continuous oxygen determination
0    other

*biomass (A-7.43)*
1*   conversion from chlorophyll data
2a*  conversion from cell counts and cell measurements
2b*  conversion from coulter counter data
0    other

### zooplankton

*species abundance (A-7.44)*
1    microscopic analysis
0    other

*stage distribution (key species only) (A-7.45)*
1    microscopic analysis
0    other

**Code   Method**

**zooplankton (cont'd)**

*individual stage weights (copepods only) (A-7.46)*
1a    dry-weight, preserved
1b    dry-weight, fresh or deep frozen samples
0     other


*biomass (A-7.47)*
1     dry-weight
2*    conversion from lengths
3*    conversion from organic carbon content
0     other


**hyperbenthos**

*species abbundance (A-7.48)*
1     visual inspection
0     other

*stage distribution (mysids only) (A-7.49)*
1     microscopic analysis
2     length-frequency distribution
0     other

*biomass (A-7.50)*
1     weighing (AFDW)
2     calculation from body-lengths
0     other

## Biological variables: sediment

**Code  Method**

### micro-phytobenthos

#### species abundance (A-7.51)
1    tissue technique
2    density gradient
0    other

#### production (A-7.52)
1    $^{14}$C tracer technique, *in situ*, bell jar
2a   $^{14}$C tracer technique, sample, *in situ*
2b   $^{14}$C tracer technique, sample, incubator
3    oxygen determination, *in situ*, bell jar
4a   oxygen determination, sample, *in situ*
4b   oxygen determination, sample, incubator
0    other

#### biomass (A-7.53)
1    C/Chl-a ratio calculation
0    other

### macro-phytobenthos

#### species abundance (A-7.54)
1    visual inspection
0    other

#### biomass (A-7.55)
1    dry-weight
2*   wet-weight
0    other

#### trace metal content (A-7.56)
1a   digestion by $HNO_3$, AAS detection
1b   digestion by $HNO_3/HClO_4$, AAS detection
1c   digestion by $HNO_3/H_2SO_4$, AAS detection
1d   digestion by other acid mixture, followed by AAS detection
2a-d  similar digestions, other instrumental detection
0    other

## Code  Method

### meiobenthos

#### species abundance (major taxa) (A-7.57)
1   microscopic analysis
0   other

#### biomass (A-7.58)
1a   dry weight, preserved samples
1b   dry weight, fresh/deep frozen samples
2a   ash-free dry weight, preserved samples
2b   ash-free dry weight, fresh/deep frozen samples
3a   blotted wet-weight, preserved samples
3b   blotted wet-weight, fresh/deep frozen samples
4*   volumetric based conversions
0   other

### macro-zoobenthos

#### species abundance (A-7.59)
1   visual inspection
0   other

#### age distribution (selected species) (A-7.60)
1   growth marks
2   frequency distribution
0   other

#### biomass (A-7.61)
1   ash-free dry weight
2   dry-weight
3   blotted wet-weight
4*   conversion from body lengths
5*   conversion from body volume
0   other

#### trace metals (A-7.62)
1a   wet tissue destruction, instrumental analysis
1b   tissue ashing, instrumental analysis
0   other

**Code  Method**

*PAHs (A-7.63)*
1a   RP-HPLC, saponification/enzymatic treatment
1b   RP-HPLC, no saponification/enzymatic treatment
2a   GC-MS, saponification/enzymatic treatment
2b   GC-MS, no saponification/enzymatic treatment
3a   ID-GC-MS, saponification/enzymatic treatment
3b   ID-GC-MS, no saponification/enzymatic treatment
0    other

*PCBs (A-7.64)*
1a   GC-ECD, saponification/enzymatic treatment
1b   GC-ECD, no saponification/enzymatic treatment
2a   GC-MS, saponification/enzymatic treatment
2b   GC-MS, no saponification/enzymatic treatment
3a   ID-GC-MS, saponification/enzymatic treatment
3b   ID-GC-MS, no saponification/enzymatic treatment
0    other

# Annex IV

# JEEP92 minimum programme

The set of sampling procedures is to be considered as a minimum programme. Many aspects will be easily incorporated in already existing programmes, and several aspects may be combined into one operation.

**Preliminary survey (not obligatory)**
(see section 4.2)

*Aim:*
to understand the general hydrodynamics of the estuary, especially the salinity distribution in relation to time and space. Other variables may be included (*e.g.* temperature, dissolved oxygen, fluorescence). The result of this survey will be the basis for the selection of benthic sampling stations.

*Sampling 1:*
vertical profiles of salinity along the entire salinity range of the estuary (CTD or ST preferred); minimal 3 samples per location (0.5 m, mid-depth and 1 m from the bottom), continuous recording is preferred.

*Sampling 2:*
at a minimum of 2-4 locations, based on expected salinity regimes, tidal variations in salinity are recorded over a full tidal cycle (13 h). This should be performed at minimum and maximum river discharge.

**Effects of tidal cycles and river discharge**
(see sections 4.3, 4.4 and 4.6)

*Aim:*
to collect samples that give an insight into the effects of tidal cycles, on a daily scale (high-low water cycle), on a fortnightly scale (spring-neap tide cycle) and on a seasonal scale (river discharge)

*Locations:*
one sampling station should be selected that falls within the salinity range $9 - 15 \times 10^{-3}$; a second option is the turbidity maximum.

*Time:*
samples should preferably be collected:
- once at the period of maximum river discharge, and once at the period of minimum river discharge;
- in the period half way between spring and neap tide (additional sampling at two subsequent spring and neap tide events is highly recommended);
- for at least 1 full tidal cycle, 13 h (25 h is preferred to detect also the diurnal cycle)
- at 1 hour intervals.

*Sampling:*
sampling at mid water depth.

**Routine sampling, water, including seston and organisms**
(see sections 4.5 and 4.6; tables 3 and 4)

*Aim:*
to collect water samples for the analysis of dissolved and particulate matter and of suspended biota, along the axis of the entire estuary;

*Locations:*
at locations that are determined by salinity of the water column, at salinities about every $3 \times 10^{-3}$ (table 4);

*Time:*
each sample should preferably be collected
- throughout the year, with higher frequencies in the spring-summer period, and dependent on the variable under consideration (table 3);
- half way between spring tide and neap tide,
- at the period half way between high and low water, sampling :
  number of replicates in tables 3 and 4, collection at mid-water depth.

**Routine sampling, sediment, including organisms**
(see sections 4.5 and 4.6; tables 3 and 4)

*Aim:*
to collect sediment samples and benthic biota at selected locations along the axis
of the estuary.

*Locations:*
at fixed locations that are determined by the ranges in salinity, and based on the
simplified Venice classification (tables 2 and 4);

*Time:*
the samples should be collected twice a year in spring and in autumn; sometimes
higher frequencies are preferred, depending on the variable/organism under
consideration (table 3);

*Sampling:*
number of replicates in tables 3 and 4.
Sampling depth is set to:
- for physico-chemical characterization of (surface) sediments the top 2 cm,
- micro-phytobenthos is sampled in the top 0.5 cm,
- meiobenthos is collected in the upper 5 cm,
- for macro-zoobenthos the top 25 cm should be sampled.

**Analytical procedures and priorities**
(see chapter 7)

A list of variables that are considered important for the JEEP92 programme is
given in the table 5 with a priority 1. Priority 2 analyses are desirable and priority 3
analyses are for specialists.

**Protocol**
(see chapters 5 and 8)

A field data inventory form (table 6) has to be completed to provide all relevant
background information on the sampling event (*e.g.* station number, sampling
time, location, tidal stage, weather conditions, information about the samples and a
list of samples that are collected).
To allow comparison and interpretation of data a JEEP92 data base has been
developed. Standard formats are therefore required. Units are given in table 5,
JEEP92 data entry (sub)codes for sampling and analysis are summarized in
Annexes II and III.

# Annex V

## Conversion factors for the calculation of biomass of:

- bacteria
- phytoplankton
- zooplankton
- meiobenthos

## Bacteria (after Bratback, 1985)

Bacterial biovolume: $1 \ \mu m^3 = 5.6 * 10^{-13}$

## Phytoplankton (after Strathmann, 1967)

For phytoplankton (other than diatoms):
$$\log C \equiv 0.866 \log V - 0.460$$
where C is the carbon per cell (in pg) and V is the cell volume (in $\mu m^3$).
For diatoms:
$$\log C \equiv 0.758 \log V - 0.422$$
where C is the carbon per cell (in pg) and V is the cell volume (in $\mu m^3$).

## Zooplankton (after Parsons *et al.*, 1984)

For mixed zooplankton samples containing more than 90% copepods the
following relation is proposed:
$$mg \ C \equiv 0.49 * mg \ (dry\text{-}weight) - 5.19$$

## Meiobenthos (after Feller & Warwick, 1988)

- Simple approximation of the volume (V, in nl) from length (L, in $\mu m$) and
  width (W, in $\mu m$):
  $$V = L * W^2 / 16.10^5$$
- Approximate conversion factors (C) to estimate the volume V (in nl) from
  the maximum width (W, in mm), total length (L, in mm) according to the
  formula:
  $$V = L * W^2 * C$$
  where C is defined:

  | taxon | C | taxon | C |
  |---|---|---|---|
  | Nematodes | 530 | Tardigrades | 614 |
  | Ostracods | 450 | Hydroids | 385 |
  | Mites | 399 | Polychaetes | 530 |
  | Kinorhynchs | 295 | Oligochaetes | 530 |
  | Turbellarians | 550 | Tanaids | 400 |
  | Gastrotrichs | 550 | Isopods | 230 |

- To convert volume (V, in nl) to dry-weight (in $\mu g$) multiply by specific
  gravity (*e.g.* 1.13) and dry-weight wet-weight ratio (*e.g.* 0.25).

# Annex VI

## Mathematical description
## of the estimation of
## weight from length regressions

(after Baskerville, 1972)

## Estimating the (mean) weight from length measurements:

1. Perform the length-weight regression according to the assumption:

$$w = c \cdot l^a \Leftrightarrow \ln(w) = \ln(c) + a \cdot \ln(l) \left.\begin{array}{c} \\ \\ \end{array}\right\} \Rightarrow \ln(w) = a \cdot \ln(l) + b$$

$$\text{say:} \quad b = \ln(c)$$

with the data

| $l_1$ | $w_1$ | $\ln(l_1)$ | $\ln(w_1)$ |
|---|---|---|---|
| $l_2$ | $w_2$ | $\ln(l_2)$ | $\ln(w_2)$ |
| ⋮ | ⋮ | ⋮ | ⋮ |
| $l_n$ | $w_n$ | $\ln(l_n)$ | $\ln(w_n)$ |

by calculating the following additions and coefficients of the linear regression Y = a.x + b where Y=ln($w$) and X=ln($l$):

$$\sum x^2 = \sum X^2 - \frac{(\sum X)^2}{n}$$

$$\sum y^2 = \sum Y^2 - \frac{(\sum Y)^2}{n}$$

$$\sum xy = \sum XY - \frac{\sum X \sum Y}{n}$$

$$\Rightarrow$$

$$a = \frac{\sum xy}{\sum x^2}$$

$$b = \frac{1}{n} \cdot (\sum Y - a \cdot \sum X)$$

$$s_{xy}^2 = \frac{1}{(n-2)} \cdot (\sum y^2 - \frac{(\sum xy)^2}{\sum x^2})$$

2. To estimate the weight from lengths, use the regression coefficients to calculate the uncorrected logarithmic weight $\hat{\mu} = \ln(w)$:

$$\hat{\mu} = a \cdot \ln(l) + b$$

then calculate the estimated weight and its variance by taking the antilog and correcting it for the use of logarithms in the regression:

$$\hat{w} = e^{(\hat{\mu} - \frac{s_{xy}^2}{2})}$$

$$\hat{\sigma}_w^2 = e^{(2 \cdot s_{xy}^2 + 2 \cdot \hat{\mu})} - e^{(s_{xy}^2 + 2 \cdot \hat{\mu})}$$

from these estimated weights the mean weight and its standard deviation can be calculated:

$$\left.\begin{array}{cc} \hat{w}_1 & \hat{\sigma}_{w_1}^2 \\ \hat{w}_2 & \hat{\sigma}_{w_2}^2 \\ \vdots & \vdots \\ \hat{w}_m & \hat{\sigma}_{w_m}^2 \end{array}\right\} \Rightarrow$$

$$\overline{w} = \frac{\sum\limits_{i=1}^{m} w_i}{m}$$

$$\text{s.d.} \overline{w} = \sqrt{\frac{\sum\limits_{i=1}^{m} \hat{\sigma}_{w_i}^2}{m^2}}$$

# INDEX